**MODEL AIRPLANES ARE DECADENT AND DEPRAVED**

# MODEL AIRPLANES
## ARE DECADENT AND DEPRAVED

## THE GLUE-SNIFFING EPIDEMIC OF THE 1960s

Thomas Aiello

NIU PRESS *DeKalb*

Northern Illinois University Press, DeKalb 60115
© 2015 by Northern Illinois University Press
All rights reserved

24 23 22 21 20 19 18 17 16 15    1 2 3 4 5

978-0-87580-724-9 (paper)
978-1-60909-178-1 (ebook)

Book and cover design by Shaun Allshouse

Library of Congress Cataloging-in-Publication Data
Aiello, Thomas, 1977-
Model airplanes are decadent and depraved : the glue-sniffing epidemic of the 1960s / Thomas Aiello.
    pages    cm
Includes bibliographical references and index.
ISBN 978-0-87580-724-9 (pbk. : alk. paper)—ISBN 978-1-60909-178-1 (ebook)
1. Glue sniffing—United States—History—20th century. 2. Youth—Drug use—United States—History—20th century. I. Title.
HV5822.G4A44 2015
362.29'63097309046
2015014044

Lyrics from "Now I Wanna Sniff Some Glue" by Joey Ramone, Johnny Ramone, Dee Dee Ramone, and Tommy Ramone reprinted with permission from Alfred Music. © 1977 WB Music Corp and Taco Tunes.

*In memory of my father*

Now I wanna sniff some glue,
Now I wanna have somethin' to do,
All the kids wanna sniff some glue,
All the kids want somethin' to do.
—*The Ramones*

# Contents

| | |
|---|---|
| LIST OF ILLUSTRATIONS | xi |
| Introduction | 3 |
| 1  The Sense of Smell | 12 |
| 2  The Mountains of Denver | 26 |
| 3  The Infection Spreads | 42 |
| 4  The Academy Responds | 56 |
| 5  The Devil in the Room | 72 |
| 6  The Glue Summit | 87 |
| 7  The Pursuit of the Monster | 104 |
| 8  The Law Down South | 121 |
| 9  The Lingering Evil | 145 |
| 10 The Children's Crusade | 158 |
| Conclusion | 173 |
| APPENDIX 1. A Description of Active Chemicals in Organic Solvents | 179 |
| APPENDIX 2. A Documentary Account of Glue-Sniffing Legislation in the Deep South | 183 |
| APPENDIX 3. Profile of Glue Sniffers across the Nation | 195 |
| NOTES | 199 |
| BIBLIOGRAPHY | 231 |
| INDEX | 245 |

## Illustrations

1. Pharmacothymia as a Form of Ego Degeneration   17
2. Childlike Regression in Male Figure Drawing   23
3. Childlike Regression in Female Figure Drawing   25
4. Volatile Solvents   46
5. Volatile Solvents by Product   48
6. Contact Trace Map   57
7. Glue Sniffing as a Function of the Number of Siblings in a Given Family   58
8. Glue Sniffing as a Deterrent to Adequate Schoolwork   59
9. Aplastic Anemia among Glue Sniffers   62
10. Ingredients of Some Airplane Glues   66
11. The Reasons Students Drop Out   82
12. Behavior and Belief among Glue Sniffers and Non-glue Sniffers   94
13. Using Segmented Variables to Predict the Possibility of Glue Use   95
14. Incidences of Reported Glue Sniffing in New York City   99
15. Glue-Sniffing Legislation   108
16. Glue Sniffing in Three Cities   109
17. City Glue-Sniffing Differentiation by Age and Sex   110
18. Organ Damage Resulting from Sniffing Glue   112
19. Juvenile Drug Arrest Rates   171

**MODEL AIRPLANES ARE DECADENT AND DEPRAVED**

# Introduction

ROOFTOP SEX ORGIES WEREN'T supposed to happen in Cobble Hill. They just weren't. The lower-middle-class Brooklyn neighborhood prided itself on peace and civility, but residents reluctantly admitted that "we are dragged down statistically by Columbia Street," the main thoroughfare of its adjoining neighborhood, Red Hook.

Red Hook had no such claims to civility. It was the boyhood home of Al Capone, a neighborhood plagued with gang activity that often spilled over into Cobble Hill. There was in Red Hook an on-again, off-again pitched battle between the local Italian gang, the Playboys, and the local Puerto Rican gang, the Black Diamonds, that often ended in violence and occasionally ended in death. But the gangs, their membership dominated by teenagers, were more than the sum of their violent parts. The Playboys and the Black Diamonds existed to use juvenile delinquency as a bonding agent for their respective communities. And this meant, among other things, the occasional rooftop sex orgy in neighboring Cobble Hill.[1]

Throughout the bulk of the 1960s, the fuel that kept such orgies in place came from an unlikely source: model airplane glue. The boys would get high and steel their resolve to get naked in front of girls and themselves. There was no specific New York law that banned rooftop sex orgies, and as of the early 1960s, there wasn't a law that banned sniffing glue, either. The city and the nation were just learning about the potential dangers of model airplane glue, through newspapers, television reports, and a flood of scientific research. Exposés about the problem gave parents vital information about what their children might be doing, but at the same time, they also let kids in on the big secret. Prior to the 1960s, it was unlikely that either the Playboys or the Black Diamonds would ever have spent their time getting high on model airplane glue. Most gang members were not enthusiastic hobbyists, coming as they did from neighborhoods where families didn't have the kind of discretionary income that would allow them to buy model airplane kits for their children. But when they learned that there was a readily available product they could use to get high, they leapt at the chance. And so did hundreds of thousands of others.

The general human obsession with inhalation for intoxication has deep roots, from the oracle at Delphi to Judaic biblical ritual. In ancient Greece, Φάρμακον meant both "drug" and "magical substance," clearly demonstrating that drugs in ancient Greece had more than medicinal effects. The Greeks wanted magic.

Nitrous oxide was discovered in 1776 by Joseph Priestley, and ether followed soon on its heels. Chloroform was discovered in 1831. As the nineteenth century became the twentieth, the development of paint thinners, varnishes, lighter fluid, polishes, and dry-cleaning supplies provided a variety of publicly available products with organic solvents that could be inhaled for some range of hallucinogenic or intoxicating effect. Model airplane glue was always one of those products, but it never appeared in any of the myriad declamations produced by those warning of the deleterious effects of such activities through the 1950s.[2]

Then, as if from nowhere, the first reports of problematic behavior with model glue appeared in 1959, when a series of children in western cities such as Tucson, Arizona, and Pueblo, Colorado, were arrested for delinquency after it was discovered they had been huffing glue. The *Denver Post* picked up on the story and did its own exposé, leading other papers to crusade in much the same way. That story, in August 1959, either provided the initial shot across the bow for research into the subject or convinced children in the area to give it a try, and over the succeeding years Colorado youth experienced a legitimate "epidemic." Police raids in Denver turned up glue sniffers everywhere. Soon youth arrested for more serious crimes like robbery were blaming their behavior on model glue.[3]

And really, coverage of the problem both spurred initial research *and* convinced kids to try it. Such was the nature of the media. Like all technologies, it shrank and expanded the country at the same time, finding stories at every edge of the nation while bringing them into every American living room. And when those stories involved the dangers associated with model airplane glue, parents looked on with horror while their children looked on with curiosity. It was that kind of coverage, and that kind of divergent reaction, that made sniffing glue one of the most unique epidemics in American history. There was no vaccine. Every victim of the disease was a willing participant. And, perhaps most important, warnings about the dangers of infection only led more people to infect themselves. It was a vicious circle: an epidemic that swelled with every new mention of the problem. It was, then, an epidemic of words more than it was an epidemic of drugs, with discussion breeding new victims every day.

A similar epidemic of words began after a 1964 series of fights on Easter Sunday in Clacton, England. Overzealous media coverage across the country reported on the violence with exaggerated headlines. Soon discussions began about the epidemic of violence among British youths. Coverage targeted Mods and Rockers, two dominant factions of English youths, for violence and vandalism. Stories were exaggerated. Some were fabricated. The three-year phenomenon led sociologist Stanley Cohen to use the term "moral panic"—an explosion of fear or broad concern about a threat from a specific source.[4] Typically that fear is exaggerated and takes on a life of its own beyond the reality of whatever threat might actually exist. When threats are culturally constructed, they only really become "problems" when there is a consensus or collective group agreement that concern is warranted.[5]

While the glue-sniffing epidemic was very different from England's teenage violence, it was certainly a moral panic. That isn't to say, of course, that the consequences weren't dire. The risks of sniffing glue were real. Toluene, the active intoxicant in model airplane glue, could cause headaches, nausea, and irregular heartbeats. It could lead to liver or bone marrow damage and caused several comas. It also created the kind of intoxicating behavior that sometimes led to accident, death, or the occasional rooftop sex orgy.

There are several elements that tend to define moral panics. First, there must be an increased level of concern, and corollary to that concern, there must, at least within one segment of society, be a functional consensus that the problem is real. The level of public concern in such situations is disproportionate to the actual threat, and the cited figures by those generating the panic are, in the words of researchers Erich Goode and Nachman Ben-Yehuda, "wildly exaggerated." Finally, moral panics are inherently temporary. Panics either become institutionalized or they disappear. They emerge suddenly and disappear just as quickly.[6]

After the glue-sniffing epidemic suddenly emerged in Colorado and Arizona, it quickly spread throughout the country. Or, perhaps, the Colorado investigations led other states to start emphasizing analysis of such behavior. And they found it all over the country. Salt Lake City's problem became national news in short order. New York's epidemic began in 1961, with health officials and law enforcement officers publicly wringing their hands about instances of glue sniffing and the overwhelming availability of a product that was, essentially, designed to be in the hands of children. In 1963, the *New York Times* recorded the city's first death, as a fourteen-year-old boy walked off his Brooklyn roof after inhaling model glue. The city's Board of Health responded by banning the sale of model airplane glue to anyone under eighteen.[7]

But it didn't help. In 1964 there were stabbings, more falls from buildings, drownings. They continued in 1965, along with more ordinances, more laws, more hand-wringing. Nothing could quarantine the epidemic. It spread throughout the country, throughout the world. In 1967 five deaths in Japan were blamed on lacquer sniffing. Stories in papers like the *Times* told of rooftop sex orgies fueled by glue-sniffing intoxication. Stories in magazines like *Time* worried over the practice, reminding its readers that the most insidious element of the new menace, and the principal reason for its swift spread, was that "it's simple and it's cheap."[8]

Marijuana had experienced much the same treatment a generation prior. It was not a dominant topic in the American mind in the early 1930s. Sixteen states had pot laws, but they were relatively innocuous and prosecutions were rare. By 1938 all forty-eight states had marijuana laws and the federal government passed the Marihuana Tax Act. There were wide-scale arrests, public denunciations in the media, and tales of the horrors of the "burning weed of hell" and the "sex-crazing drug menace." There are obvious parallels here to glue sniffing thirty years later, but in the case of marijuana, not only was the drug relatively illicit even before the panic but, more important, the campaign against it was fueled by a moral crusade from the Federal Bureau of Narcotics, which wanted pot under its jurisdiction and fed stories to the media and submitted sample legislation to states and the federal government to create the concern that brought the drug under its control. There was deviance and intoxication. There was a health concern and a moral concern. There was a brief explosion of concern that abated relatively quickly after the laws had taken their place in the statute books. But it was a concern that was completely created by what Howard Becker called "moral entrepreneurs"—authors and organizers of the concern for a specific self-interested reason. Glue sniffing never had that element.[9]

But it did have the media to drive the panic. Moral panics can begin because of genuine sentiment or because one of the actors has something to gain. They can start from the top down or the bottom up, or from representatives and groups in the middle—like, for example, the media. The problem with that interpretation, however, is that organizational entities like the media don't have a moral status. There must be a preexisting latent fear to foster the development of a moral panic, but that fear must find a directed expression by a moral agent and must be carried by a facilitating entity such as the media or politicians. Or, in the words of Goode and Ben-Yehuda, "All the organizational efforts in the world cannot create public concern where none exists to begin with. At the same time, concern needs an appropriate triggering device and a vehicle to express itself in a moral panic, and for that, interest group formation and activity are central."[10]

Model airplane glue had several triggering devices, chief among them parents.[11] Or, perhaps, the triggers of families, local law enforcement, and civic groups all had parents as common members. As did the media. As did politicians. The glue-sniffing epidemic was fundamentally different from other public concerns because of the age of the potential sniffers and the good intentions of both the product and the dealers. John Springhall has traced moral panics relating to youth to the first half of the nineteenth century, and he argues that such panics come at times when society is further forced to confront modernity, when panics "induced by fears of new technology [interact] with revised forms of popular culture." And with the progression of huffing various solvents from Joseph Priestley's nitrous oxide in the eighteenth century to model airplane glue in the twentieth, that analysis seems to be accurate.[12] Still, the fact that parents, the chief moral drivers of the panic, were among every group that participated in the development of the epidemic, certainly drove the rhetoric.

That rhetoric carried the epidemic along, but at the same time, the rhetoric had to be there. Ignoring the problem wouldn't make it go away. The rhetoric was dire, calamitous. Heroin, too, was a big problem, but it affected adults. Model airplane glue affected adolescents, with the majority of offenders falling between the ages of eight and sixteen. The death toll would rise into the hundreds, newspapers claimed the sky was falling, and researchers sought answers to the reasons for the deaths and the long-term health effects lingering for those who survived.

The epidemic started in 1959, pushed by those early arrests in Colorado and Arizona, but the notion that kids in Colorado invented the practice, that no one had ever thought to sniff glue before 1959, seemed inherently far-fetched. The long history of gasoline and solvent sniffing in the United States was there for anyone to see. At the same time, kids didn't stop using model airplane glue and other household solvents after the 1960s came to a close. Instead, the nation's focus would drift from adolescent glue sniffing to the countercultural student movement, with its attendant devotion to marijuana and psychotropic drugs. Pushed as it was by opposition to Vietnam, southern race policies, and bureaucracy in general, the movement (and its drug use) came to embody a tumultuous era fraught with violence, civil disobedience, and massive sea changes in American life and law. Glue sniffing faded by comparison.

And so the glue-sniffing epidemic began after glue sniffing started, and it ended well before adolescents stopped sniffing. In that way, too, it was unique among American epidemics. To that end, the story of the abuse of model airplane glue begins well before the onset of the "disease," with American psychiatric thinking about the sense of smell and its potential to induce or signify a disturbing psychological makeup. From the 1920s to the 1950s, Freud held sway in this regard, and consequently such thinking tended, in the tradition of Freud, to equate abnormal devotion to the sense of smell with sexual deviance. Freud also tied juvenile delinquency to a child's relationship to the parents, wherein troubled living situations led to a malformation of the ego or superego, and this ultimately led to the kind of masochism that included huffing model airplane glue. Those psychological debates are examined in chapter 1, carrying American attitudes about smell and sniffing through the 1950s.

The story of the epidemic begins in chapter 2, as glue sniffing is discovered, debated, and researched from 1959 to 1962. The reactions in the popular press to the early onset of the problem and the growing number of users populate chapter 3, followed by the academic response to such reports. Juvenile court judges, doctors, scientists, law enforcement officers, and social workers all lent their own perspectives to the epidemic, trying to figure out the most likely candidates for infection, the most likely health problems they would face, and the most likely prevention methods that would actually work. The Hobby Industry Association of America (HIAA) got involved, too, understandably worried that its member corporations would suffer under the weight of the hysteria. For all of their talking, however, the problem only continued to metastasize around them, with more examples of abuse, more examples of crimes committed while children were under the influence of model glue, and more examples of untimely deaths. Beginning in 1959, researchers held annual conferences to parse out the intricacies of the menace, but their 1964 meeting in Denver, the epidemic's ground zero, was by far the most significant. By 1964 there was an overwhelmingly large sample set of both cases and case studies, leaving the glue summit to evaluate the past problems and prepare for future manifestations. Their work continued throughout the rest of the decade, as researchers stubbornly butted heads over the profile of the common user, the health implications of use, and the best methods of rehabilitation. At the same time, state and municipal legislators worked diligently to curb the problem through legislation. It was a curious idea. Model airplane glue was designed to provide children with helpful, character-building ways to occupy their time, but as fear of its abuse permeated communities, lawmakers found themselves with no choice but to regulate the product. They did so in different ways, depending on location, population size, extent of the problem, and length of time since the epidemic had arrived. The South got around to glue regulation later than most regions, not because southern states didn't have a similar glue-sniffing problem as everyone else but because legislators had their hands full fighting desegregation mandates and other outgrowths of the civil rights movement. When they did get around to it, however, southern states would enact a series of laws more comprehensive than any other region in the country. Those southern states are described in chapter 8 to demonstrate how such laws were passed and to demonstrate the different kinds of legislative responses to the problem—responses that ultimately paralleled the various laws passed in cities and states across the country and North America. The laws were helpful in giving police the authority to respond to glue sniffers and the crimes they committed while under the influence, and they almost certainly raised the stakes for kids considering the practice and warned them away because of the potential for punishment. But they didn't stop the epidemic. It continually expanded, even though new laws appeared on city and state books every year of the 1960s.

Finally, the Testor Corporation, one of the founding members of the HIAA and the most prosperous of the model airplane glue companies, developed an irritant that kept sniffers from turning to its products. It didn't end the problem, but it helped to end the hysteria. By that time, however, the counterculture and its antecedents had taken their place on the American landscape, largely drowning out the worried cries of parents frustrated by their children's abuse of glue and other household solvents. And other solvents were important. When kids found that glue wasn't really an option anymore, they turned to household cleaners, gasoline, aerosol sprays, and even the nonstick cooking aid Pam. They still do. For that reason, Hawaii congresswoman Patsy Mink led one final children's crusade in the early 1970s, arguing for a national law to regulate such solvents. State laws were never as effective as federal statutes. Didn't civil rights prove that?

Her effort, however, failed. The glue irritant was in place, parents now worried that pot and LSD were as common as Snickers bars, and the "epidemic" was over. Had that epidemic been strictly the result of model airplane glue, Mink's law would almost certainly have passed. But it wasn't. It was the result of the public outcry about model airplane glue, and that was something fundamentally different. Whether or not children were still dying from inhaling toxic solvents (and they were), the hysteria had dissipated, and this convinced Congress that legislation was beside the point.

Glue sniffing was not a constituent component of the counterculture. Its chief participants were younger than thirteen, well before the age of susceptibility to a military draft that their older counterparts found immoral at best.

Although sniffing glue sometimes became a communal activity, it tended to be practiced in solitude, without pained recitations about the injustices of bureaucracy, Jim Crow, and Vietnam. This isn't to say, however, that there was no connection at all between the two phenomena. Both emphasized drug use, of course. Both caused a concerned backlash from American parents, and ultimately the backlash drawing on public fear of the counterculture would supplant the fear of model airplane glue, as manufacturers added an irritant to their adhesives and Richard Nixon began trumpeting his devotion to "law and order." The crusade against glue, in fact, would provide a road map for those crusading against hallucinogens, amphetamines, and marijuana. Finally, and perhaps most important, there was a connection between the glue-sniffing epidemic and the drug use of the counterculture in the feeling of alienation that drove them both.[13] "Students," the Students for a Democratic Society (SDS) declared in their 1962 Port Huron Statement, "are breaking the crust of apathy and overcoming the inner alienation that remain the defining characteristics of American college life." It was one of five different times that the statement mentioned alienation, and though preteens didn't have the eloquence of Tom Hayden, they too demonstrated a clear sense of alienation from their schools, their parents, and the larger world around them. For some, it was poverty and difference that drove such alienation. For others it was what Emile Durkheim called the "anomie of affluence." Regardless, that disconnect from the understood norms of society created a social distance between users and those around them. SDS argued that the alienated society resulted from the inability of people "to turn their resources fully to the issues that concern them."[14] The group wasn't thinking about children when they crafted their statement, but such a conclusion certainly applied to them. "The developmental discontinuities in our ways of raising children," wrote Kenneth Keniston in the early 1960s, "produce, to greater or lesser degree, inner conflicts between the dependency needs exploited in childhood and the independence required in adulthood."[15] Thus the drug use associated with the counterculture fed from the glue-sniffing epidemic, which started earlier, skewed younger, and set the precedent for the paranoid backlash that would follow.

Mink's quixotic last-ditch effort to regulate glue sniffing was a fitting mirror for the entire epidemic. Law enforcement officials, parents, teachers, researchers, and legislators all took up the fight against the monster of model airplane glue, never fully understanding that each element of the fight made the monster stronger. That being the case, when they finally stopped fighting as much, they were able to tell themselves that the monster had grown smaller. The epidemic must have been over. It was a self-fulfilling prophecy, a blind spot in

American thinking that created a problem through scrutiny then ultimately solved it through inattention. Children were still dying from inhalation of household solvents in 1980, but there was no epidemic. This made it much easier for Lloyd Bridges to tell his fellow air traffic controllers that he had "picked the wrong week to stop sniffing glue."[16] It was funny, and *Airplane!* was released a full decade after the epidemic had subsided. Besides, if the hysterical response to model airplane glue was built on every new hysterical response, then all you had to do was close your eyes and pretend it wasn't there. It was almost as if glue-sniffing rooftop sex orgies never happened.

The hysteria that developed around glue sniffing would set a precedent for the modern War on Drugs, a new moralistic assault on drug use that would begin with the presidency of Richard Nixon, who would institute his own drug panic in the early 1970s, expanding the federal drug budget and creating new agencies to police drugs. He was able to succeed largely because of his ability to create a sense of paranoia among those who had been conditioned by panics like the glue-sniffing epidemic. Ronald Reagan learned the same lessons, using the drug panic in the 1980s to pass two new major pieces of legislation in 1986 and 1988 and grow an even larger budget for the drug war. And wars, of course, do not exist to treat victims. They exist to take prisoners, and most of those prisoners—the vast majority of them nonviolent offenders—would be black. And so the War on Drugs that both followed and flowed from the fear generated by the glue-sniffing epidemic also left a long racial legacy, creating, as Michelle Alexander has noted, a permanent underclass and a new version of segregation. Each drug panic built upon the previous panic, creating more legislation, more institutionalization, more resources. It also set attitudes about the morality or deviant nature of drug use.[17]

So moral panics are short-term phenomena that have long-term consequences. Moral panics redraw the lines of morality and social acceptability in society (as well as legislative lines and lines of corporate responsibility) and are, therefore, a "long-term social process rather than [a] separate, discrete, time-bound episode."[18] That said, while the glue-sniffing epidemic produced legislation and corporate change that created functional long-term change, it was ultimately different than the marijuana panic that came before it and the drug panics that followed.

This is the story of a decade of open eyes, widened by fear. It is the story of a harried response to the troubling realization that adults had been supplying their children with the tools of their demise. Parents could lock their doors. They could cautiously peer out of their windows in fear. But the contagion was in the house. It was on top of dressers and under Christmas trees. The only thing left was panic.

# The Sense of Smell

WHEN ABRAHAM ARDEN BRILL, one of the leading American psychologists of the early twentieth century, stepped to the podium at the annual meeting of the American Psychoanalytic Association in late December 1930, he was well aware that the psychological consequences of the sense of smell had long been seen as ancillary to other sensory stimuli: sight, sound, touch. His audience at the New York Academy of Medicine had spent years and volumes parsing out the intricacies of the tactile, the audible, the visual, as they related to neuroses and psychoses of all shapes and sizes. Smell played little or no role in the life of the most evolved of the species. "As a rule," he told his audience, "civilized man is not only independent of this sense, but dislikes any odors emanating from human beings."[1] And if olfactory independence was the mark of civilization, then it was only natural that any emphasis on smelling was therefore under the purview of "animals, primitives, and even semi-enlightened people."[2] Smell was imperative for the survival and mating practices of mammals. The South American jungle Indians, the South Sea Islanders, the Möis of Indochina, the Eskimos, all had a highly developed sense of smell. Moving up the chain of civilization, even Asians, southern Europeans, and South Americans—each falling somewhere between Eskimos and honest-to-god white people on the scale of social evolution—developed cultures where the sense of smell played a role.[3]

This being the case, any predominant olfactory tendencies by civilized white people had to be considered aberrant. British psychologist Havelock Ellis had argued as early as 1906 that "there is a special tendency to the association of olfactory hallucinations with sexual manifestations." Nineteenth-century sexologists like Germany's Richard von Krafft-Ebing and Ireland's Connolly Norman had said much the same thing. Swiss psychiatrist Eugen Bleuler, most famous for coining the term "schizophrenia," dealt with smells and their hallucinations in his seminal *Lehrbuch de Psychiatrie*, arguing that smell hallucinations were associated with the last stages of manic paresis and schizophrenic delusions.[4]

And then there was Freud. The Austrian luminary argued that the repression of the pleasure of smell in civilized adults played a direct role in certain predispositions toward nervous diseases, leading specifically to sexual repression, "for we have long known what an intimate relation exists in the animal organization between the sexual impulse and the function of the olfactory organ." Freud was saying much the same thing in 1908 that psychologists had been saying for much of the late nineteenth century. Sex and smell were undeniably, biologically related. The following year, Freud appeared to have refined his thinking, arguing that "psychoanalysis has filled up the gap in the understanding of fetichisms by showing that the selection of the fetich depends on a coprophilic smell-desire which has been lost through repression."[5] Still, it didn't require much of an argumentative leap to conclude that, if the sense of smell was linked to sexual problems and fetishes, smell itself could be fetishized.[6]

Brill was nothing if not familiar with the literature and wrung his hands over the nose and its relation to sexual development. "It is quite obvious," he argued, "that normal sexual development must depend on the existence of an unimpaired sensorium."[7] In particular, Brill argued that defective sexual development appeared most commonly as "feminine masochism," a baroque and ostracizing term for male homosexuality. But women weren't immune to such deficiencies. Brill referenced the work of German otolaryngologist Wilhelm Fliess, who argued that women had *genitalstellen*, or genital spots, in their noses, which could be treated with cocaine to ease painful menstruation.[8]

But Brill had his own case studies to bolster his claim that an emphasis on sniffing and smelling was directly related to psychosis.[9] As his audience looked on, he told them about patients who couldn't respond sexually to women who smelled even vaguely like their mothers, about a patient who could only attain ejaculation by rubbing his testicles and smelling his hands. About "anal-sadism." About mysophobia. About foot and shoe fetishes. About dementia praecox with the dominant symptom of smearing feces on the wall to be surrounded by the foul odor. About a patient who longed for dead carcasses because only the dead wouldn't push away when he tried to hug, kiss, and smell them. "Without going any further," Brill noted in a moment of false discretion, "we can say that this patient's great need was to be able to wallow in the slimy carrion of some dead female body. He was not only coprophilic, but also coprophagic; he often devoured horse manure and on occasion his own excrements."[10]

Such had always been the case. Napoleon had an overdeveloped sense of smell, as did Cardinal Richelieu, Henry IV, and even the notorious American socialite Clara Ward, who made a name for herself in the 1890s by marrying a Belgian prince and spent the rest of her days referring to herself as the Princesse de Caraman-Chimay. Oscar Wilde had discussed such smell attractions in *The Picture of Dorian Gray*, as had Emile Zola, Ben Jonson, and Somerset Maugham. Baudelaire, Poe, Shelley, Nietzsche, Tolstoy. All of them fretted over the particulars of smell. But Brill left no stone unturned. God himself had problematic smell disorders before humans became further civilized. "Jehovah was particularly fond of the sweet savours of burnt offerings—an odor which one would hardly relish nowadays."[11] His children slowly began to refrain from such gifts as humanity developed further and thus learned to repress such problematic relationships with odors.[12]

And that was the thing. "Cultured humanity has a greater capacity to repress sexual smells than any other feelings. . . . Olfaction is endowed with greater affectivity than any of the other sensory activities despite their common origins." It was a catch-22 for humanity. Repression of olfactory tendencies was a hallmark of civilization and thus understandable and, for lack of a better word, good. But at the same time, that repression, like all repressions, could double back and cause significant neurotic disorders, particularly sexual disorders. "When man assumed an erect posture and turned his nose away from the earth, smell fell more or less into disuse, which increased with the advance of primitive civilization, when the first sexual taboos were established," Brill argued. "But, as the first taboos were primarily of an incestuous nature and smell was still an active sex function, primitive man was under special stress to curb this sense."[13] For all of its other problems, that repression ultimately brought us out of the ooze, made us civilized, and kept us from the uncivilized horrors of the South Sea Islands or the Arctic Eskimo population.

Of course, there was one group among the truly white and civilized that had yet to fully develop, a group that was, in constitution, far more primitive than adults: children. Smell gradually declined as children aged, or perhaps it moved back from the conscious mind to the subconscious. But until that transition was complete, children were always going to be fundamentally at risk. Their emphasis on odors put them closer to the primitives than their civilized parents. Journalist Gare Hambridge used a questionnaire to rank the most pleasing odors for adults, and the overwhelming winner among men was the scent of pine. Pine was second among women's favorite smells. And pine, Brill commented, "has, of course, a marked resemblance to that of turpentine."[14] The Hungarian psychoanalyst Sándor Ferenczi had noted as early as 1916 that odors of turpentine, gasoline, and other chemicals led directly to the "sublimation path of anal-eroticism,"[15] and such fondnesses surely began in childhood, as children showed a marked tendency toward the smell of turpentine, gasoline, and similar solvents. The adult push for civilization had placed the children of properly cultured parents in the center of a dangerous game.[16]

The stakes were clearly high. Civilized white people might be able to look laughingly at the olfactory tendencies of animals, shaking their heads as dogs sniffed each other's asses. They might be able to raise their noses in judgment at the Third World and its inhabitants and use smell as another problematic justification for the latter's inferiority and lack of social evolution. But they couldn't stomach either of those reactions when it came to their children. They actually *cared* about white kids. And white kids were at risk.

Between October 1931 and July 1933, researchers in San Antonio, Texas, recorded seventy-two cases of gasoline and kerosene poisoning in children ranging from ten months to four years old. The children weren't sniffing it. They were drinking it. But the aroma suckered them in. In one case, a seventeen-month-old boy drank an ounce of kerosene. Doctors tried a gastric lavage. They tried hypodermic atropine and caffeine sodiobenzoate. Nothing seemed to work. Two hours after the child's admission, his temperature had risen to 106.5, and an hour after that he died. Another boy drank gasoline and died less than two hours later. Of course, not all of the cases were fatal. Overall, 28 percent of gasoline patients died, 9.2 percent of kerosene patients, but all suffered dramatic painful symptoms. Still, "the toxicity produced by the ingestion of coal oil or gasoline need not cause the great concern that aspiration or inhalation of these hydrocarbons causes." The prognosis became markedly darker when the patients aspirated the products along with ingesting them, which created pneumonitis in the lungs. But breathing the fumes of the petroleum-based products also produced disorders of the nervous system ranging from cyanosis to convulsions to coma. "We feel," the researchers argued, "that the toxic fractions of these oils reach the vital centers of the central nervous system much more rapidly and in larger amounts when absorption takes place in the lungs. Absorption of these substances from the gastrointestinal tract alone is much slower."[17]

Again smell had proved dangerous, particularly to those who had yet to develop. It was the attack on the nervous system, and the respiratory center motor areas in particular, that turned out to be so troublesome. And the most dangerous aspect of such poisonings wasn't through the mouth. "The prognosis may be said to be in direct ratio to the amount of the hydrocarbon that enters the lungs."[18] The aroma drew children to the toxic chemicals, and whether they chose to drink the chemicals or not, it was the clarion call of that smell that ultimately felled victims.

Of course, if gasoline was a rare commodity then such cases would be a curiosity more than anything else, but the danger that came from such findings was heightened by the fact that gasoline was becoming more and more common (a phenomenon that would, one generation later, push inhalation of glue into the forefront of the national consciousness). Americans were using a billion more gallons of gasoline in 1940 than they had in 1927, pushed by the rise of the automobile. In the late 1930s, the number of gasoline pumps grew at an annual rate of a hundred thousand. While researchers continued to chronicle that growth, however, the public never adopted the widespread fear of narcotic abuse that they would in later generations. As mentioned in the introduction, the paranoia of the 1930s emphasized marijuana.[19]

Still, there's no way that any researcher wringing his hands over the potential threat of gasoline, kerosene, or turpentine could possibly interpret the poisoning of those under four years of age as drug use, as intentional, as deviant. It was intent that somehow compromised the white bastion of civilization. "The psychoanalytic study of the problem of addiction," argued Hungarian psychologist Sándor Radó in 1932, "begins with the recognition of the fact that not the toxic agent, but the impulse to use it, makes an addict of a given individual."[20] Radó emphasized that cravings for any chemical or narcotic were all forms of one single disease, a disease Radó called "pharmacothymia."[21]

Users of any narcotic chemical did so either to prevent pain or to generate pleasure, both falling under the broader cope of what Radó called "the pharmacogenic pleasure-effect."[22] But this effect was ultimately temporary and led in the long term to depression and self-destruction.[23]

A child was a simple megalomaniac, "radiant with self-esteem, full of belief in the omnipotence of its wishes, of its thoughts, gestures, and words."[24] But as experience mounted, as disappointments began to pile up, the ego was reduced (again the looming influence of Freud) and narcissism continued to prove problematic. "Or we might put it, the ego must make over its psychology from that of a supercilious parasite into that of a well adjusted self-sustaining creature."[25] So, in adults, addiction created a cycle of depression and elation based largely upon the desire to maintain that original narcissism. The consequences for children, then, were even more pronounced, because their narcissism had yet to be shoved into the background by a fully developed ego.[26]

In adults such intoxication ultimately led to a moral crisis, where the ego would interpret this bodily harm in service to temporary pleasure with either recriminating self-reproach or full-on displacement, wherein the addict blamed (in the thinking of 1930s psychology) his "genital organ." Here again appears the concept of such behavior being childlike. "Primitive thought finds displacements of this sort very easy," Radó argued. "We often hear small children say: 'I didn't do it. My hand did it.' The life of primitive peoples is replete with instances of this sort."[27] And so the addict regressed back to the narcissism of childhood. What would happen to children who were already living in that regressive state? And were turpentine and gasoline really the kinds of products that could cause such problems?

Yes. The composition of gasoline varied based on the region where it was refined, but all varieties included paraffins, olefins, naphthenes, and aromatic

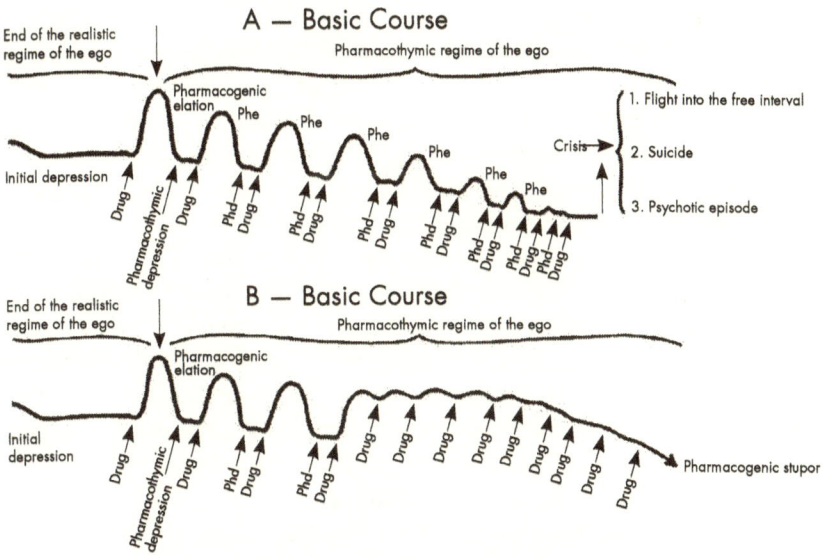

**SCHEMATIZED COURSE OF PHARMACOTHYMIA**

PHARMACOTHYMIA AS A FORM OF EGO DEGENERATION. Drugs, in Radó's formulation, reacted with the brain to create a state of mental declension that, if prompted by crisis, could lead to interminable consequences. Source: Sándor Radó, "The Psychoanalysis of Pharmacothymia (Drug Addiction)," *Psychoanalytic Quarterly* 2, no. 1 (1933): 22.

hydrocarbons, among them several kinds of xylenes, benzene, and, importantly, toluene (the dominant aromatic in model airplane glue, which would cause so many problems in the decades to come). Gasoline also included tetraethyl lead, which created lead poisoning problems largely unrelated to respiration (and ultimately led to the later dominance of unleaded gasoline), but absorption of gasoline through the respiratory tract was the principal problem. Hyperemia, petechial hemorrhages, subpleural extravasations, and "as in any acute general intoxication, the types and degree of pathologic change are influenced by the interval between ingestion or exposure and death."[28] William Machle, an MD from Cincinnati writing in the early 1940s, was clearly concerned, but his emphasis remained on accidental exposure, pleading the case of petroleum workers, dry cleaners, and others who risked relatively constant exposure to such volatile chemicals.[29]

One of the most volatile of those chemicals was toluene ($C_6H_5CH_3$), a hydrocarbon also known as methylbenzene or toluol. It boiled at 110.4 degrees Celsius. It mixed readily with alcohol, chloroform, carbon disulfide, ether, and petroleum benzene. It was insoluble in water. Toluene could dissolve iodine, sulfur, and phosphorus. But perhaps even more importantly, toluene dried quickly, and therefore was a readily used chemical in rubber, lacquer, munitions, and glue.[30]

Throughout the 1940s, researchers publicly worried about exposure to toluene, just as they had for gasoline. Toluene poisoning was principally caused through respiration, although as for gasoline, absorption through the skin could also prove problematic. Also like gasoline, production was rising. In 1930, just over sixteen million gallons of toluene were produced. In 1939, that number rose to twenty-four million.[31]

Researchers studied painters and gauged their exposure to primer, lacquer, and paint, noting that average exposure to toluene ranged from 100 to 1,100 parts per million. The dominant symptoms shown by the painters were perforated nasal septums and enlarged livers. Such were abnormalities, but the researchers were hard-pressed to call them symptoms. There was, they discovered, a surprising absence of severe illness. But there was still no consensus. Another group studied workers in the munitions industry (toluene was used as starting material for trinitrotoluene, an active explosives ingredient). Workers there experienced fatigue, nausea, headaches, and impaired motor functions, all at parts per million comparable with the painter study.[32]

Finally, a third study tracked the progress of workers in "a large industrial plant." Toluene fumes in the plant acted as a depressant and attacked the central nervous systems of employees. It was an irritant to the mucous membrane. This "toluene intoxication" was undeniably toxic, and patients ultimately received blood transfusions, iron, calcium, phosphorus, and other extensive vitamin therapies to aid their recovery.[33] But whether the prognosis was encouraging or frightful, it was clear that toluene could be just as problematic as gasoline, and unlike gasoline, its use continued to grow through multiple industries. Still, though the researchers were divided on the kinds of problems exposure to toluene could cause, they were united on the reality that it did cause problems. Whether you worked with paint, lacquer, rubber, explosives, or glue, you were a waiting potential victim. And just as early gasoline poisoning studies had done, the toluene worries of the 1940s solely emphasized accidental exposure.

But as the 1940s became the 1950s, the focus shifted to intentional exposure.[34] The country was growing up, shedding its childhood narcissism, and venturing into the world of ego reduction, disappointment, and depression. Intentional inhalation and all the incumbent problems of addiction became paramount.

Psychologist J. M. Schneck had, in the mid-1940s, reported a case of chloroform addiction that led to a schizophrenic condition, but with that exception, the scientific literature had been quiet on the subject of intentional use. In 1951, however, Orris W. Clinger and Nelson A. Johnson first analyzed intentional huffing—in this case the active benzene in gasoline—"deliberately indulged in for the fantasy-provoking, hallucinatory effect." And as the psychoanalytic community of the late nineteenth and early twentieth centuries predicted, the case studies of Clinger and Johnson were all adolescent boys.[35]

Clinger and Johnson worked at the Warren State Hospital in Warren, Pennsylvania, and they told the story of "a Negro boy of 16" who was admitted to the hospital in 1946. The boy attended an integrated high school and didn't appear to have any problems with his white classmates. In fact, some of his best friends were low-income whites from a nearby neighborhood. His father owned a garage, and as a youngster the boy watched as his older brother and his friends would huff gasoline from the cans around the shop. It wasn't a long-standing habit with the older boys, but every kid wants to be like his big brother. Thus began a long relationship with gasoline fumes that ultimately spiraled into addiction. "If I was doing something else, like playing baseball, I wouldn't think about the gas and it didn't bother me," the boy admitted, "but when I didn't have anything else to do, I just got possessed with the idea that I had to smell gasoline—and then I couldn't rest until I got some."[36]

The boy was relatively unpopular in school. His grades in elementary school were very good, but they had become progressively worse as the years progressed. He was a loner. He was dissatisfied with the way his parents punished him. Gasoline became his outlet, the hallucinatory effects of inhaling the fumes his only moments of pure joy. "I would see people dressed up in pretty clothes walking along the streets and talking to each other," he reported. "I could tell that they were happy and gay and enjoying themselves. Then I could see lots of water and many different bright colors."[37] Such would be a profile all too familiar in the years to come, when cases like this teenager's would become increasingly sensationalized as part of a larger epidemic.[38]

And there was, again as predicted, a strong deviant sexual component to such behavior. The boy would go out to the woods with his friend Joe, and the two would huff gasoline before masturbating together. There were at least nine other boys who participated in the practice. The gas fumes would often bring erection, and ejaculation would come through "sexual fantasies, sometimes of a heterosexual nature, but more often of a homosexual type, involving boys of his own age."[39]

As the boy's inhalation tendencies turned from habit to addiction, however, Joe too would fall away. Sniffing became a solitary practice, making the boy coarse and angry to his family, even more withdrawn at school. Finally, his parents believed they had no choice but to send him to Warren State. There his schizophrenic tendencies were patiently documented. He was tested with every known apparatus available. But his parents became frustrated with the slow course of treatment and pulled him from hospital care.[40]

Back home, the problem only became worse. While he was generally even more secluded than before, he would insert himself into situations for attention, attempting to answer questions in class even though he had no grasp on the correct answer. He began to write obscene notes to white girls in his class and try to fondle them when he could. "When any white girl would pass him on the street he would invariably turn to this friend and say, 'I'd like to rape her.'"[41] Meanwhile his hallucinatory tendencies increased, as did his angry behavior toward his family. They eventually had no choice but to readmit the boy to the hospital, where he would stay for almost two years.[42]

Away from the temptations of gasoline, the boy began to thrive. His hallucinations dissipated, his willingness to associate with others increased, as did his intellectual and emotional intelligence. After his second release, he was able to take a job at his father's garage, engage in a successful romantic relationship with a girl, and repair his relationships with his family members. This was not psychiatric illness causing a predilection toward sniffing gasoline. This was gasoline disrupting the constancy and flow of the brain, gasoline *causing* psychiatric illness. Still, as successful as incarceration was at ending the torture of addiction, "the Negro boy" wasn't the only case of gasoline-related hospitalization, and there were clearly others in his peer group taking part who were never admitted to the hospital. "This practice," concluded Clinger and Johnson, "may be more prevalent than heretofore known."[43]

This was the kind of language that would endorse a panic over model airplane glue a generation later, so it is unsurprising that in the aftermath of such rhetoric, the cases seemed to keep coming. In 1948, an eleven-year-old white boy was admitted to a University of Minnesota hospital after a weeks-long bout of gasoline sniffing. Like his counterpart from 1946, the boy was generally healthy and had healthy parents, but his home life was problematic. His parents' violent arguments eventually led to divorce. A grandparent had died. The boy had been sniffing gas since the age of five, but only sporadically. After the divorce, his use became sustained, and his nervousness and depression ultimately led his mother to admit him to the hospital. Under a regimen of sodium amytal, the boy admitted to frightening hallucinations under the influence of gasoline, of ugly little men "who came out of the ground and who threatened him with death." But such didn't scare him away from the practice. In fact, the little men originally conspired with the boy to remove his father. "Father heckled me and gave my mother and me heck at everything we did," he confessed. "I want to make my father go away, so I can go off someplace with mom and we can make a living together and be happy together on a farm."[44] Again, his behavior, mental health, and physical condition all improved after hospitalization and the enforced removal from gas that came with it.[45]

But as studies of adolescent addiction evolved through the 1950s, the assumptions of researchers appeared just as ethnocentric as they had in the 1930s (or, for that matter, in the 1800s). "Adolescent opiate addicts are preponderantly of Negro or Puerto Rican origin," concluded medical researchers Donald Gerard and Conan Kornetsky in 1955, which was not only "in accord with the ecologic findings of others but also lends weight to hypotheses which stress socially-conditioned frustrations and special processes or interactions among groups of adolescents as the more basic etiologic factors of adolescent opiate addiction."[46] But this wasn't an argument for primitivism. Black and Puerto Rican adolescents were subject to constant discrimination and prejudice, leading inevitably to problems with self-esteem and a glass ceiling on expectations for success in the adult future. Interestingly, however, the same study concluded that the majority of adolescent opiate addicts came from stable middle-class backgrounds, a group that demonstrated far more deviant behavior than the lower classes.[47]

Gerard and Kornetsky's study of admitted adolescents to the US Public Health Service Hospital in Lexington, Kentucky, emphasized psychiatric causes for such addictive instincts. Along with Rorschach tests and other common psychological grading scales, the researchers also asked the patients to draw human figures. The patients then had to tell a story about each drawing. The drawings were judged on their descriptive ability, the level of childlike regression in the images, and the rigidity or animation of each. Researchers assumed that patients would draw childlike images with little psychosexual differentiation, but while the drawings of the addicts were more rigid than those of the control group, and while there was a low number of incidences of adequate psychosexual differentiation, the number of "normal" drawings remained pretty close. The addicts were more likely to be schizophrenic, the controls were more likely to be "normal." But the conclusion from this data that "youths living in urban areas where illicit opiate use is widespread do not become addicted independently of psychiatric pathology" seemed hopelessly flawed.[48] Everyone in the control group and every addict studied was either black or Puerto Rican. All came from similar neighborhoods either in Kentucky or New York City. Socioeconomic status didn't even prove to be the factor the researchers originally assumed. And so the Gerard and Kornetsky study actually demonstrated the opposite of the authors' intent. Opiate addiction tended to *cause* psychiatric problems, it affected adolescents of *all* ethnicities and classes. Although this pseudoscientific emphasis on race as a determinant for illicit addiction would continue into the nation's later obsession with glue, victims of addiction could come from anywhere, regardless of their ability to render accurate human forms.[49]

And the numbers of addicts were growing. Throughout the 1950s, there was an annual average of five hundred new cases of adolescent male narcotics abuse in New York City alone, the vast majority of them involving heroin. A study produced by New York University's Research Center for Human Relations argued that such use flourished in the poorest neighborhoods of the city and often led to criminal activity. But it didn't always lead to addiction. In one study of ninety-four heroin users, less than half used the drug regularly. Still, whether use was the result of addiction or recreation, all instances had some common factors that generated a crucible in which users were created. There was, for example, a delinquent subculture that provided encouragement, validation, and (most important) drugs to create such habits. There were personality disturbances coming from family problems, fear of the responsibilities of coming adulthood, or a general lack of trust in one's surroundings. But like Gerard and Kornetsky, the NYU project claimed that socioeconomic status couldn't be considered a constant factor in such use. The lower classes produced an overwhelming number of delinquency cases, but drug use was something different. It could affect male adolescents from every class, and then the level of that drug use would condition any further delinquent behavior, as it ultimately became a way of life for the user, an all-encompassing

Body Image Regression Together with Inadequate Psychosexual Differentiation in a Drawing of a Male Figure

Body image regression:
- (a) head wider than trunk;
- (b) geometric forms—the round head, square neck, rectangular legs

Inadequate psychosexual differentiation:
- (a) no distinguishing clothing;
- (b) no distinguishing sexual characteristics in head or face, bosom, hips or genitalia

CHILDLIKE REGRESSION IN MALE FIGURE DRAWING: Gerard and Kornetsky sought to understand the state of the patient's mind through such drawings, hoping to see how each gauged human development. Source: Donald Gerard and Conan Kornetsky, "Adolescent Opiate Addiction," *Psychiatric Quarterly* 29 (July 1955): 469.

cope over every other thought and action. And there was, as of the late 1950s, very little available help for these juvenile addicts and users. There just wasn't enough knowledge of the problem. Psychiatry could make inroads into curbing the problem, but "users do not easily 'take' to psychotherapy."[50] In addition, addiction wasn't something cured quickly. It required sustained treatment over vast swaths of time, which required money, discretion, and a lack of stigma for something that was sure not to remain private.[51]

Faced with these treatment challenges, psychologists and researchers looked on helplessly, without proven treatments, without a consensus about the genesis of the problem, and with a clear knowledge that there was a vast preponderance of narcotic agents that were far easier to find than heroin. If those same adolescent males, for example, discovered that toluene and other benzene hydrocarbons were available in products they had in their bedrooms, would they really need to participate in the kinds of activities required of regular heroin users? Or, if those adolescents discovered the magic of getting high in their bedrooms with such chemicals, would that make them seek out new highs like heroin? Every answer seemed to produce a new question, and with the progression of the problem from the 1930s to the 1950s, it seemed to researchers that they were on the precipice of a legitimate epidemic. Significantly, however, the American public never developed a commensurate panic. The literature's emphasis on lower-class and minority use, combined with the exoticism of heroin to the majority of American families, stunted the momentum of a broader movement against such opiates.

Not to mention that, as the incidents of addiction and narcotic use grew and as the cases of gasoline poisoning and toluene poisoning began to mount, the diagnoses and conclusions remained the same. Researchers emphasized that addiction problems were metastasizing all around them, but they remained mired in a state of clinical stagnation. In 1959, for example, a New York psychologist named Paul Friedman considered the sense of smell in modern life. He quoted Freud, Wilhelm Fliess, Ferenczi, von Krafft-Ebing, Connolly Norman. He discussed Emile Zola and quoted Marcel Proust.[52] A generation had passed since Abraham Brill mounted the podium at the American Psychoanalytic Association meeting, but commentators were still saying the exact same things about the relationship between smell and deviance, smell and sex, smell and addiction. But by the end of the 1950s, there was far more addiction, far more sex, and far more deviance than there was in the early 1930s. Such was the state of psychology as those problems—deviance, sex, smell, addiction—threatened to combine in a new form, a form potentially more dangerous and certainly more available than other supposed threats to adolescent American males and the families that worried for them: glue.

THE SENSE OF SMELL   25

## Inadequate Psychosexual Differentiation in a Drawing of a Female Figure

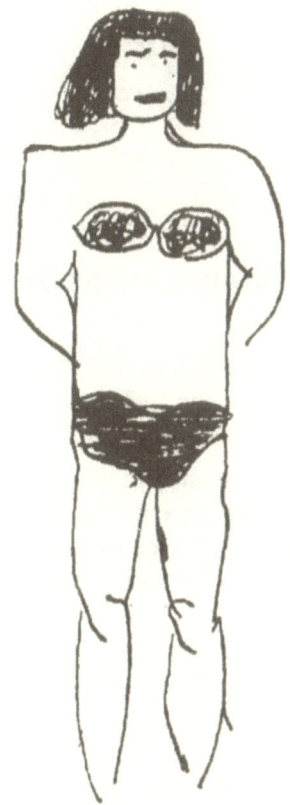

Inadequate psychosexual differentiation:
   (a) body contour masculine (note broad shoulders, narrow hips, muscular bulges in arms and legs).
   (b) lack of feminization of the head and face, which except for the exaggerated but nonaesthetic treatment of the hair could serve as well for the head of a male figure

CHILDLIKE REGRESSION IN FEMALE FIGURE DRAWING: Again Gerard and Kornetsky demonstrated an inadequate understanding of human development in child psychiatric patients. Source: Donald Gerard and Conan Kornetsky, "Adolescent Opiate Addiction," *Psychiatric Quarterly* 29 (July 1955): 470.

# The Mountains of Denver

THERE IS A MAJESTY about mountains that encircle a city, giving everyone in their surroundings a sense of place, a broader sense of self that reaches up into their climbs and returns with identity and purpose. At the same time, though, the mountains dwarf everything around them, a visual reminder to those staring up at their larger peaks that humanity is small, hopelessly small, and therefore susceptible to the whims of predators not cautioned by monumental rocks jutting up from the earth. The Rocky Mountains, for example, stare down over the people and buildings of Denver with watchful eyes, their snowy caps and imposing structure trying to brace everyone against threats from the west. But there are so many holes, so many problems that could seep through the jagged rocks and make their way to the population below.

Pueblo, Colorado, is 120 miles south of Denver down Highway 87, an old boomtown on the Arkansas River once home to sheriff Bat Masterson. Shortly after Masterson's reign, Woodrow Wilson came to Pueblo on September 25, 1919, to stump for the League of Nations. He collapsed after his speech and had a stroke one week later, incapacitating him and functionally ending his active presidency. Far from the snow-capped peaks of Denver, Pueblo suffered through hot dry summers. By the sweltering summer of 1959, Bat Masterson was dead and the rugged West had been tamed. There were no more presidents stopping by. The Colorado State Fair, held every year in Pueblo, was still months away. Children were bored. They were stuck inside to avoid the relentless heat. And it was there in those air-conditioned homes that the children finally found something to occupy their time. They decided to "spread liquid glue on the palms of their hands, then cup their hands over their mouth and nose and inhale deeply." It was a "quick bang," a "mild jag," and ultimately, it was something to do. "A variation of this is to soak a handkerchief with glue and hold it over the mouth and nose." No one was really sure when the habit started, but in the hot summer of 1959, Pueblo police discovered it and began arresting children. So did police in Tucson, Arizona, another lonely city boiling in the summer sun.[1]

Lots of cities were lonely and boiling in the summer sun. Boredom was a phenomenon that reached beyond the bounds of Colorado and Arizona. Declamations of the epidemic that would appear throughout the 1960s cited Pueblo and Tucson as the points of genesis for the contagion, but none of these accounts provided any evidence of specific environmental factors related to those places that would spur a set of unique acts. This is because they were not unique acts. Pueblo and Tucson were points of genesis for the epidemic not because the act of children sniffing glue was exclusive to those cities. The history of the childhood obsession with sniffing and the availability of model airplane glue made the act relatively common and, if nothing else, a fait accompli. Those cities were the points of genesis for the epidemic because police decided to make arrests, and the press decided to report on them.

Denver was the largest city in a rural state. The population in 1959 hovered around 490,000. It had grown exponentially from 1940 to 1950, and again from 1950 to 1960, but the growth slowed through the decade of the glue-sniffing epidemic, expanding by roughly 20,000. The population of Pueblo, 120 miles to the south, followed the same pattern in microcosm, its approximately 90,000 residents in 1959 growing by more than 6,000 over the course of the next decade. The difference between the two is that Pueblo's suburban population was incredibly small, almost 95 percent of its residents living in the city. Denver's statistical area, however, was well over one million people, more than half its population living in the suburbs surrounding the city. Unsurprisingly, metropolitan Denver had greater incidences of poverty. More than 117,000 city residents lived in census tracts with a poverty rate of 20 percent or more, and almost 40,000 of those were Hispanic. Of the 154,000 children living in census tracts classified by alternate poverty rates, almost 25,000 lived below the poverty level.[2]

Tucson was much smaller than Denver, with its statistical area around 225,000 residents in 1959, growing to more than 290,000 over the course of the 1960s. In both decades, the vast majority of residents resided within the city limits. Only around 10 percent of Tucson-area residents lived in the suburbs around the city. There was a significant Hispanic population, like there was in Denver, as Tucson's county shared a border with Mexico.[3] Pueblo, Tucson, and Denver each had a population that was potentially susceptible to the spread of an abuse phenomenon for a cheap available substance such as model airplane glue, but there wasn't a pattern that made that susceptibility apparent. There weren't any shocking deviations from the norm that made such places radically unique.

There also weren't any laws against sniffing model airplane glue, but police arrested the kids anyway. It seemed dangerous. It seemed detrimental. Officials around the country were fretting about the potential dangers of all kinds of new toxins that people were using to get high. While the children of Colorado and Arizona were experimenting with glue, for example, the CIA was at Stanford testing the effects of LSD and other psychotropic drugs, all of which were also still legal. Model airplane glue, however, was different. The kids being discovered in Colorado and Arizona weren't in college, they weren't even high school upperclassmen. They were children. And they didn't have to seek out a CIA program or willing beatnik to get their drugs. Their parents had purchased the drugs, willingly, assuming that making model airplanes was a healthy and constructive method to while away hot summer afternoons.[4]

Spurred on by the knowledge that "police in Pueblo, Colo., and several other cities in the West and Midwest" had come in contact with such behavior, the *Denver Post* decided to investigate and became the first major American newspaper to report on the problem. Its August exposé included only one example of model glue intoxication, and it was an example of accidental intoxication. A boy working tirelessly at a model airplane in a room without ventilation was overcome by glue fumes. Still, the initial shot across the bow in the assault on model airplane glue would set a standard for much of the coverage that followed throughout the next decade. The newspaper listed the principal culprits of intoxication and bodily harm existent in various formulas of model glue: "chloroform, carbon tetrachloride, acetone, ethylene di-chloride, toluene (or toluol), xylene (or xylol), isopropyl alcohol." It reported on an interview with Samuel Johnson, director of Denver's Poison Control Center, who warned of long-term harm caused by inhalation, principally bone marrow problems, anemia, and liver damage. It sounded a dire alarm from law enforcement. "I'd rather see a kid go on a drunk than go on a jag with this stuff," said one juvenile official. "If you think a person who has been on a big drunk looks bad you should see some of these kids. They have bags under their eyes, and the whites of their eyes are greenish-yellow." Finally, it suggested ways to make the problem go away, settling on putting the onus on manufacturers to label their products to warn of serious injury from accidental or intentional inhalation.[5]

Such would be the formula used by countless journalists, researchers, and politicians in the coming years. It fanned public fears by using experts to provide extreme examples of problematic health or behavior. But resting in the spaces between such coverage was an ominous reality. Even though the *Post* article didn't provide any examples of intentional inhalation, intentional inhalation had surely begun in Denver. The juvenile official had known exactly what the glue sniffer looked like. He had seen them. And the dire alarm undergirding his statement clearly hinted that he was worried that the problem would grow.

It did. Less than a year after the *Post*'s initial exposé, in June 1960, the Denver police department's juvenile bureau announced that it had investigated close to fifty glue-sniffing cases in the first half of the year. The problem was no longer a curiosity. In the span of several months, it had become an epidemic. "This practice is extremely dangerous," said Philip Gartland, head of the juvenile bureau, "and a kid can die from it if he gets too much." Joe Moomaw, head of the city's crime lab, explained that inhalation carried benzene into the respiratory system, where it evaporated and deposited in the body. He explained that the effect was similar to that of a mild anesthetic, dulling the senses and producing pleasing sensations. But those deposits weren't worth the high. "I'm afraid some kid will get hold of too much of that stuff, and we'll have a fatal case on our hands."[6]

It was a frightening premonition, and one that would come true at various times as the epidemic spread, but the question remained: Why were kids in Denver, so far from the sleepy towns of Pueblo and Tucson, doing this? There were, in essence, two answers, neither of them completely satisfactory. First, there was a legitimate possibility that Denver adolescents had been sniffing glue at the same time as their fellows to the south, and that the realization by Denver law enforcement that such activities were taking place ultimately led to such investigations. With such a large sample size in a metropolis like Denver, however, that answer seemed lacking. It was clear in the 1959 exposé that officials had seen such cases before, but it seemed fundamentally implausible that police officers would ramp up their vigilance against the practice because of a brief article in the Sunday magazine of the local paper. The other answer was that, although Denver had seen cases of glue abuse before, the exposure that the practice received locally and the hyperbolic rhetoric with which it was described acted as a generative agent, convincing children who never knew they had access to such intoxication that it would be easy to give it a try. Assuming that the notion of a spontaneous, coincidental explosion of such use was impossible, the best conclusion seems to lie somewhere in the space between the two potential answers. Police were obviously going to be more vigilant about a problem once it began receiving publicity, and kids were obviously going to be curious about the nature of model airplane glue. It was a perfect media-generated storm, and everyone in Denver knew that if the police had come across fifty cases, there were myriad others that had gone unnoticed by law enforcement. This isn't to say that the media was intentionally manufacturing unnecessary concern. There was genuine worry about young children abusing this substance, from newspaper editors and the experts they consulted for their stories.[7]

By the summer of 1960, the word was spreading. Cases appeared in Scottsdale, Arizona, and (significantly) Los Angeles, California. The *Journal of the American Medical Association* (JAMA) weighed in on the new epidemic, arguing that "it was inevitable that the more imaginative of our teen-agers would also discover the exhilarating and intoxicating properties" of model airplane glue, which didn't mean, of course, that fighting the problem wasn't necessary. "These materials are liver poisons, and the frequent inhalation of concentrations sufficient to cause a jag are decidedly harmful to this organ and may cause severe damage to the respiratory and circulatory systems."[8] It was a short article, an editorial comment in answer to a subscriber's question about the practice, but it was the opening salvo in what would become the American medical and scientific community's dogged research program to map the physiological, psychological, and criminological weight of the problem.

The criminological element of glue sniffing was significant. Even at the outset of the epidemic, glue was seen by many as an aid to more problematic outgrowths of juvenile delinquency. It was also seen as a form of juvenile delinquency itself.

Of course, while there was little available research on sniffing glue in 1960, there was plenty of delinquency analysis. "I would there were no age between ten and three-and-twenty," wrote Shakespeare in *The Winter's Tale*, "or that youth would sleep out the rest; for there is nothing in the between but getting wenches with child, wronging the anciently, stealing, fighting." His message was clear: nothing good could come from adolescence. And twentieth-century psychologists tended to agree. As early as 1945, Austrian-born psychologist Fritz Redl distinguished four types of juvenile delinquents. First, there was the child who was ostensibly normal but reacted negatively to poor parenting, troubling living situations, or specific trauma. Then there was the child whose inevitable period of "adolescent turmoil" ultimately led to delinquent behavior. Then there were the neurotics. Finally, there was the "genuine delinquent," who suffered from some kind of malformation of the ego or superego.[9]

It was this easy to make the argumentative leap back to Freud, which meant that such delinquents were in dangerous peril from, if not defined by, "anal sadism," masochistic tendencies, and "passive-feminine strivings." And every social factor that helped trigger such actions mobilized, as Melitta Schmideberg noted, parroting the father of the discipline, "certain psychological reactions which stem from infantile conflicts."[10] For Schmideberg, delinquency was a paradoxical denial of and submission to masochism. Adolescents were stressed. They were scared. They were confused, overwhelmed, and each of those crippling emotions came at the very time that kids were seeking experimentation and freedom.[11]

It was a significant problem, but it wasn't only a problem for adolescents. As Los Angeles physician Alfred E. Coodley argued, "preadolescence comes in for its share also."[12] During preadolescence, established standards and values begin to fade, conscience and insight begin to disappear. It is during preadolescence that children lose their sense of identification with the culture of adults. "The parent becomes a representative of the value system of society as opposed to those of children."[13] Ultimately, such children respond by forming cliques with their peers, groups that are far more permissive and structurally opposed to the morality of parents and other adults. "After all," wrote Coodley, "youngsters need some place where preadolescent traits can be exercised and even tolerated."[14] And they weren't getting it at home. Coodley noted "limited frustration tolerance, ability to sublimate, or capacity to delay immediate impulse satisfaction" as principal preadolescent frustrations.[15] Failure led to rage, success led to hubris. The code of the delinquent was clear: "I do not need you or depend on you or love you; I am omnipotent; I hate you. You, the adult, are malevolent, so I can fight you without guilt."[16]

The most prominent and frightening outgrowth of juvenile delinquency was drug use, which led potentially to drug addiction. There were many ways to define addiction. Users could develop a physical dependence, they could demonstrate a tolerance to the substance that required them to use more and more to achieve the same effect, or they could develop a psychological dependence that, divorced from physical cravings, led users to believe that the quality of their lives would diminish without the drug. The problem with this analysis, for Coodley, was a problem that researchers would debate throughout the decade. Model airplane glue might not be a drug, and addiction in juveniles didn't seem to be a legitimate problem. "Even among those teenagers who have allegedly been addicted," he argued, "the withdrawal syndrome is mild to absent." Far more universal was consistent experimentation with narcotic substances, from glue to heroin, and if any real addiction to such toxins existed among adolescents, it was psychological, a method for coping with the paradoxes of youth, the desire for freedom and the reality of adult control. That being the case, "the treatment of such delinquents demands maximum flexibility." Therapists needed to be available at all times by telephone, they needed to call delinquents on their lies, but most important, they needed to realize that their efforts were glorified (though therapeutic) stall tactics. "The question as to whether delinquents are treatable cannot be answered in a categoric fashion," Coodley explained. "So much depends on the degree of ego and superego disturbance, on the environmental setting, and on the emotional and physical capacities of the adults and contemporaries in coping with these disturbed youngsters."[17]

Delinquency, though none of the psychologists were mentioning it, also depended on the easy availability of outlets for antisocial behavior. Coodley's analysis appeared in June 1961. The following month, two child psychologists at the University of Minnesota presented a series of case studies featuring adolescent gasoline addiction. They demonstrated a picture of childhood inhalation that would become all too familiar. In the first case, a fifteen-year-old boy who had bounced around from foster home to foster home developed an extensive gas-huffing habit over a period of six months. At some times, he reported abject fright at "seeing jelly-like figures that were formless and flabby-shaped," but at others, there was "an erotic component" that would produce "transient erections." In another case, a thirteen-year-old boy from a broken home lived with "a hostilely-seductive mother"[18] and spent much of this time grabbing her breasts. In other moments he was sniffing gasoline, a habit that had continued for two years. If he couldn't find gasoline, "the boy inhaled carbon monoxide from the exhaust pipes of cars."[19] Finally, a twelve-year-old girl with "an extremely domineering" mother and a passive, alcoholic father had been sniffing gasoline at least five times a week for more than two years, until finally complaints of "dizzy spells, eye difficulty, vague pains in the neck and back, and having queer sensations in her head"[20] drove her to the hospital. She was the only of the three whose extended stay in the hospital didn't seem to diminish her desire to sniff. "The pleasure and gratification from inhalation were greater than the benefits of giving it up."[21]

All three had family problems. Two of the three had an absent father. There was instability, emotional upheaval, and a general ambivalence, if not outright rejection, on the part of parents toward the children. There was also, researchers noticed, a marked need for approval among the patients, but this need could manifest itself in different ways. Some juveniles were aggressive and caustic, hoping to win attention through poor behavior. Others were meek and "overcompliant," a technique used to hang on to what few positive attentions they managed to receive. Most demonstrated some combination of both, pragmatically changing tactics as the need arose. In the end, however, the researchers came to the conclusion that gasoline-sniffing children "are attempting to obtain certain erotic gratifications, obtain feelings of security, and maintain their self-esteem simultaneously."[22] The same could be said—would be said—of those sniffing model airplane glue.

Many were, in fact, already saying it. Back in Denver, public fear about the problem escalated when seven teenagers, ranging from thirteen to seventeen, were arrested near the Curtis Park Recreation Center for sniffing glue in October 1961. It was becoming "a major problem among juveniles in Denver," said Phil Gartland, quoted in the *Denver Post*, further fanning the flames of public concern. Joe Moomaw explained that the high came from a diminished oxygen supply in the blood, which could produce "giddiness and dizziness" but could also cause brain damage and death. "We are averaging about thirty boys a month now on this glue-sniffing problem," reported Harold Jones, the Denver juvenile court's chief probation officer. "This is not supposed to be habit-forming from the standpoint of drug addiction, but some of the kids tell us that they just can't get rid of the habit." Even those who wanted to stop were telling officials that they couldn't. Was it a physical addiction? Probably not. But this didn't make the psychological belief in dependence any less real. Even more problematic, Jones argued, was the fact that the spread of the problem wasn't centering on seventeen-year-olds. It was centering on those in junior high and "even into the grade schools." The illicit use of model airplane glue might have been growing, but so too was the panic.[23]

Tom Adams, director of the Lookout Mountain School for Boys, a reform school in nearby Golden, Colorado, admitted that at least fifty of his charges were there for sniffing glue. The problem just continued to grow in Denver. It was still there in Pueblo, too. "I understand that Pueblo authorities are so concerned that they are working out agreements with storeowners in an effort to prevent the sale of this type of glue to juveniles." It was a novel idea, and one that would eventually take prominence among American legislators, but there was—as of October 1961—no specific statute either to force hobby stores to comply or to punish those children who were abusing model airplane glue. Instead, Colorado officials charged users with a more generic juvenile delinquency. They had to do something. "Last week," said Tom Adams, "we received a boy who told us he had been using about eight tubes a day. We had a problem in getting him back to a normal routine of eating and sleeping."[24]

With every such report, more adults were made aware of the epidemic, more hysteria gripped the population, and more children learned how to get high. With numbers reaching thirty prosecuted cases every month in Denver alone, the consequences of covering glue-sniffing arrests were easily worth the more pressing need to keep parents and officials vigilant about the problem. Besides, if publicity didn't come from scientists and law enforcement officers, it was surely going to come from other places. It was the simple start to a self-fulfilling prophecy. And then it got bigger.

"There were kids, eight, nine years old, who were sniffing airplane glue to get high on," said Lenny Bruce, staring out mischievously into the laughing crowd. "These kids were responsible for turning musicians on to a lot of things they never knew about, actually." By 1961 Lenny Bruce was at the height of his powers, after scraping by as a comedian in the late 1940s. His stand-up albums, his appearances on *The Steve Allen Show*, and his unrelenting, often vulgar, commentary on current events and the existential crises of the age had made him famous in the 1950s. On February 3, 1961, he gave a legendary performance at New York's Carnegie Hall. Later that year, his album *American* appeared, a collection of stand-up performances from 1960. True to form, in one short bit, Bruce was able to circumscribe the fear and danger associated with glue sniffing. "So, I had a fantasy, how it happened," he told his audience. "Kid is alone in his room. It's Saturday. The kid is played by George Macready:

> "Well let's see now, I'm all alone in the room and it's Saturday. I'll make an airplane! That's what I'll do. I'll make a Lancaster, a good structural design. I'll get the balsa wood here, cut it out, sand it off. Now a little airplane glue. I'll rub it on the rag, and uh . . . (sniff) . . . Hey now! Ha ha. Oh, I'm getting loaded. Is this possible? Loaded on airplane glue? Maybe it's stuffy, and I'll call my dog over. Filicka! Filicka, come here darling and smell this rag. Smell it, you freaky little doggy! Smell the rag, Filicka. Filicka! Filicka! He's up there! I've done it! I'm the Louis Pasteur of junkiedom! Out of my skull for ten cents. Well, there's much work to be done now. Horses' hooves to melt down. Noses to get ready."
>
> Cut to the toy store, any toy store, any neighborhood. Kid walks in, "Hello, Schindler. Nice store you got here. Give me, eh, a nickel's worth of pencils. Big boy tablet. Some ju-ju beans. 'Tailspin Tommy' book . . . and two thousand tubes of airplane glue."[25]

In October 1961, as Denver was worrying frantically over the glue-sniffing epidemic plaguing the city, Bruce would be arrested for obscenity in San Francisco, in the first of a long line of obscenity arrests that would ultimately be his undoing.[26] But Bruce's popularity was at its height at the time of *American*'s release, and his bit ensured that the story of the glue-sniffing epidemic would reach into hundreds of thousands of new homes.

Surely no one was pacing the aisles of toy stores, waiting to subtly purchase two thousand tubes of airplane glue, but the bit pointed out the principal cause for official concern. Model airplane glue was cheap and easy for anyone to buy. Colorado was listening. In December 1961, state representative Ben Klein announced his plan to propose three new pieces of legislation to curb the state's growing epidemic. One would make selling model glue to minors without parental permission a misdemeanor. Another would make it illegal to sniff glue. A final law would revise the motor vehicle code to prohibit driving under the influence. Each would carry a one-thousand-dollar possible fine and a maximum of six months in jail. Children caught sniffing glue would be remanded to the jurisdiction of the juvenile court. "There are no laws on the books now to cope with this ever-increasing problem," Klein worried. The move almost certainly carried an element of opportunism, as politicians rushed to demonstrate the most legislative concern about model glue. But the Denver congressman was also an attorney and a former juvenile court probation officer and had seen the problem up close. He reminded his constituents that several juvenile delinquents had been charged with car theft and other more dangerous crimes as a result of sniffing glue. Concern for the welfare of children was inherently important, but so too was concern for the welfare of everyone else. It wasn't a victimless crime.[27]

Klein's legislation would ultimately fail at the state level, though Denver would pass its own municipal ordinance that mirrored his concerns. But with the problem beginning to spread across the country, many wondered if the federal government could be any help. The only attempt by Washington to regulate potentially dangerous household solvents was the Federal Hazardous Substances Act of 1960. The law required labeling on any toxic substance, "if such substance or mixture of substances may cause substantial personal injury or substantial illness during or as a proximate result of any customary or reasonably foreseeable handling or use, including reasonably foreseeable ingestion by children."[28] It defined as toxic any substance that could cause illness through "ingestion, inhalation, or absorption through any body surface."[29] The problem, of course, was that sniffing model airplane glue was neither customary nor reasonably foreseeable. Children were most certainly inhaling it, but they did so beyond the prescribed use of the product, giving the HIAA a reasonable claim against contamination warnings on their glues. And even if they had included such warnings, product labels would be wholly ineffective in stopping the spread of the glue-sniffing epidemic because children who huffed glue did so precisely because of its toxic nature. If anything, the use of such labeling would only alert more children to the reality that they could misuse the product for narcotic effect. It was, ultimately, the same catch-22 that hampered media coverage of the epidemic.

Meanwhile the media coverage was becoming more dire. Back in Denver, reports surfaced that juvenile delinquency had risen by 18.2 percent in 1961, a spike that officials attributed almost entirely to the glue-sniffing craze. In fact, glue sniffing itself increased 190 percent in 1961 from the previous year, jumping from 95 to 276 arrests. Many of those glue sniffers found themselves in the hands of authorities for other crimes such as burglary and truancy. "Something," said Denver juvenile court judge Philip Gilliam, "must be done to control the problem."[30] Juvenile authorities in the city contacted the US legislature and the Colorado legislature, begging that something be done to control sales of glue to young people. Klein's state law never passed. The Hazardous Substances Act didn't apply. And Denver children continued to fall victim to the epidemic. Leo Gomma, a juvenile probation officer, tried to remain positive. "A lot can be done through informing the public—maybe more than through legislation—to control the problem." That positivity, however, seemed problematic, as every indication pointed to public information as a principal culprit in spreading the disease. And no amount of information was going to make model airplane glue any less accessible, any less affordable for kids of any age. "Sure it may not be good for you," admitted one user, "but boy when you can get stoned for a dime, that's something."[31]

Then again, if it was just lying around in your parents' garage, gasoline could get you stoned for nothing. In early 1962, another gasoline-sniffing study appeared that added to the growing profile of potential sniffers. "All gasoline addicts so far reported have been boys,"[32] explained William M. Easson—falsely. Easson, from Saskatchewan's University Hospital, missed the twelve-year-old Minnesota girl, but his general conclusion that the vast majority of sniffers were boys was correct. He, too, pointed out a preponderance of absent or distant father figures, or fathers who "alternated sadistic brutality towards their sons with over-permissiveness."[33] The boys came from "unsettled" homes, to be sure. They also seemed to have overindulgent mothers, a childish self-centeredness, and a marked emotional distance from those around them. Faced for the first time with anxiety and frustration, with situations beyond their ability to control the outcome, the boys turned to an easily available substance to relieve an otherwise unbearable tension. Gasoline provided euphoria and hallucinations. It provided an outlet. And while the children seemed to evince a real psychological dependency on the substance, hospitalization (and therefore separation from any available gasoline) didn't provide them with any significant withdrawal symptoms.[34] It was a profile that would become very familiar to those researching model airplane glue. All that was left was for someone to come along and provide similar case studies on sniffers of something other than gasoline. It was only a matter of time.

No one would have to wait long. Several months after the appearance of the Easson report, in July 1962, *JAMA* published the first in-depth study of glue sniffing among children. Helen Glaser and Oliver Massengale worked in the adolescent clinic and department of pediatrics at the University of Colorado Medical Center in Denver. Their place in the epicenter of the growing epidemic put them in close contact with the glue-sniffing phenomenon and provided all the case studies they could want. The study they produced would become the seminal work in the field, grounding all future research and providing a foundation for everyone's understanding of the problem. "Deliberate inhalation of solvent vapors from plastic cement to induce sensations of euphoria and exhilaration has recently become such a widespread practice among older children and adolescents that it is a source of major concern to physicians as well as to school and juvenile authorities," the study opened. Such cements and airplane glues were the principal components of hobby kits, and therefore easily accessible to children. Children sometimes put glue onto a rag, a handkerchief, or a piece of gauze, before holding it in close contact with the nose. Sometimes they squeezed the glue into a paper bag. Sometimes they put it in a pan, heating it to create a "more rapid and concentrated vaporization of the cement." Of course, the problem with paper bags and pans was that they weren't subtle. They couldn't be used at school, for example.[35]

Huffing at public places like school was going to be untenable soon enough anyway. Glaser and Massengale noted a significant tolerance buildup in chronic users. Achieving the intoxicating effects of model airplane glue, which initially took just a few sniffs, could ultimately require as many as five tubes to achieve comparative results. This wasn't evidence of physiological addiction, they argued, but sniffing glue did seem, in many cases, to accompany severe habituation. "The immediate effect is one of pleasant exhilaration, euphoria, and excitement, closely simulating the early effects of alcohol." Then there was often diplopia, tinnitus. "Drowsiness, stupor, and unconsciousness then may ensue, the children remaining unresponsive for periods as long as an hour or more."[36]

The costs of such intoxication, however, could be high, ranging from bad breath to nausea to anorexia. Glue sniffers were irritable and inattentive in school, sometimes losing consciousness in class. And the habituation that came with chronic use convinced many to steal the glue when needed. Drugstores, toy stores, hobby shops, and supermarkets all carried it, and it was small enough to shove discreetly into a pants pocket or jacket. Although it seemed clear that sniffing glue could be problematic in the long term, "little is actually known about possible damage to organ systems resulting from deliberate inhalation of cement vapors." Inhalation tests on laboratory rats and rabbits demonstrated an "increased weight of the spleen and kidneys and caused degeneration of the seminiferous tubules of the testes." The tests used benzene and alkylated benzenes like toluene and xylene, all principal ingredients in various model airplane glues, but didn't yield the kinds of results that could be conclusive in assessing eventual human health risks from extended, chronic use. Besides, aromatic hydrocarbons like benzene, toluene, and xylene weren't the only volatile ingredients in organic solvents like model glue. Glaser and Massengale's report added a significant number of potential culprits to the *Denver Post*'s original list. There were the aromatic hydrocarbons, along with the halogenated hydrocarbons of chloroform, carbon tetrachloride, and ethylene dichloride. But there was so much more to worry about. Many glues included ketones such as acetone, cyclohexanone, methyl ethyl ketone, and methyl isobutyl ketone. Then there were the esters (amyl acetate, butyl acetate, ethyl acetate, and tricresyl phosphate) and alcohols (butyl alcohol, ethyl alcohol, and isopropyl alcohol). Many used methyl Cellosolve acetate. Many used hexane. All could get you high. All were volatile organic solvents. And all could do serious damage. Understanding the dangers was extremely important, argued Glaser and Massengale, because the hysteria that had developed in response to the rapid spread of the epidemic left people with false fears and misunderstandings about model airplane glue. "Many children," they explained, "share the so far unsubstantiated belief that glue-sniffing produces insanity and death, and we know of one case in which glue was first used by an adolescent boy in an attempt to commit suicide."[37]

Glaser and Massengale presented a series of case studies. There was the loner eleven-year-old from a broken home with a six-month habit. There was the fifteen-year-old boy with a psychotic alcoholic mother. Another fifteen-year-old "engaged in several glue orgies with his friends."[38] Three more Latino boys from impoverished broken homes demonstrated poor grades in school, truancy, theft, and other forms of accompanying juvenile delinquency. It was a bleak portrait, but the studies allowed Glaser and Massengale to modify the gasoline-sniffing portrait to describe the would-be glue sniffer. Although their numbers for Denver glue-sniffing arrests were low, they were able to describe the exponential growth in the practice beginning in 1959. "In the Denver area it is considered by responsible juvenile authorities to be the most serious problem they face."[39] The Denver sniffers—at least the ones who made it to the hospital or the juvenile courts—were as young as seven and as old as seventeen, with an average age of thirteen. As the gasoline profile demonstrated, glue sniffing seemed to be a decidedly male problem, with only six girls arrested among the hundreds of Denver cases. In addition, 80 percent of the arrests were Latino boys, most of them poor, many of them from broken homes. Many were repeat offenders. Thirteen had "older siblings who at some time were picked up for the same offense."[40] Only eighty-five boys were arrested solely for glue sniffing; the rest had also committed robbery, truancy, and other similar juvenile offenses. "Many of these children were undoubtedly emotionally disturbed," noted Glaser and Massengale, "and one of the many questions to be answered has to do with cause and effect relationships between chronic use of glue and psychiatric disorders."[41]

As Eric C. Schneider has noted in his groundbreaking study of the American relationship with heroin, impoverished neighborhoods did not serve simply as the locus of crime but as the creator of crime. Violent crime in New York City, for example, skyrocketed in the 1960s, but it didn't grow equally. Robbery tended to pit a perpetrator race against a victim race, for example, leading many whites to blame blacks and Latinos for the profusion of crime and the drugs (like heroin) that drove them. While such assumptions increased racism within the city, they also belied the fact that impoverished neighborhoods, created by the late ruptures of the Great Migration and the simultaneous migration of service jobs back to the increasingly aggressive Sunbelt South, were populated overwhelmingly by minorities. Realty discrimination and the centralized location of public housing also played a role, as did substandard minimum wage work. These conditions created high dropout rates for students (and high truancy rates for those who stuck around) and put those former students—mostly young minority males—on the streets.[42] Such drove the development of black and Latino crime and drug use, but the same variables created a disaffected group of younger children who would ultimately turn to glue. While glue affected all races and classes of young males, its minority hold in the cities followed familiar patterns.

In turn-of-the-century concerns over the growth in opium abuse in the shadow of the late Industrial Revolution during the Gilded Age, commentators depicted the problem as an "Asiatic vice," tying the use of the drug to foreignness and difference.[43] When cocaine entered the public discourse as a panic of its own in the early twentieth century, pundits attached a racial component to that fear, transferring the nineteenth- and eighteenth-century white fears of black sexual animalism and an inability to control more primal urges to black drug use. That was why black men were such a threat to white women. It was why they were such a threat to themselves. And that was why racial control was so necessary. Early examples like this demonstrate that even in the early phases of the public's interpretation of drug use, race subsumed the more realistic factors of class, urbanization, and demographic shift in the profusion of drug abuse.[44]

As Joel Fort noted in 1969, "the ethnocentrism or provincialism indulged in, or implicitly accepted, by most of mankind helps to give a narrowness of perspective in regard to the drug issue."[45] But Glaser and Massengale's concern over the cause-and-effect relationship between glue use and behavior, despite racial conclusions, was necessary. It was a chicken-and-egg scenario that researchers would continue to grapple with over the next decade. Did poor grades and other forms of juvenile delinquency serve as warning signs of potential use, or did sniffing glue lead to poor grades and delinquent behavior? In the early stages of the epidemic, answers to such questions were unclear. Small sample sizes and scattered reports made conclusions impossible to draw. But as the problem spread out from its epicenter across the nation, infecting urban and rural communities in every part of the country, arguments about which came first seemed beside the point. With each new public warning about the epidemic, more children were infected, because public warnings were themselves the principal contagion. And if publicity really was the generative agent for sniffing model airplane glue, the stakes set by the Glaser and Massengale study were about to be raised.

*Consumer Reports* was one of the country's most popular magazines. Founded in 1936 and published by the nonprofit Consumers Union, it was dedicated to researching products and helping customers make wise purchases based on sound science and testing. It also, however, reported on the potential misuse of certain products, on their potential dangers to people who already owned them. And in January 1963, the magazine published its own exposé on the growing menace of sniffing glue. "Glue-sniffing, one of the latest of adolescent crazes, unlike such harmless idiocies as telephone-booth stuffing, may do physical and psychological harm to its practitioners." The *Consumer Reports* article traversed much the same ground trod by Glaser and Massengale, using their study as the basis for its own comments. The author rehearsed the potential damage to users and explained the trouble in producing adequate evaluations of model airplane glue because "different manufacturers use a variety of combinations of volatile solvents in their glues and cements." Of course, the other hindrance to more knowledge about the possible dangers of such substances was that "no one has thoroughly investigated the subject because the ordinary uses of the volatile solvents involved rarely if ever produce such concentrated doses as a glue-sniffer gets."[46]

While the specifics of harm were hard to gauge, however, the simple fact of harm was all too apparent. The Massachusetts Department of Public Health, for example, called for the public to help them erase the problem after a teenager died from sniffing glue. Public health officials urged parents to take their children to the doctor immediately if they suspected possible sniffing. They encouraged employees of hobby shops and toy stores to contact the police "if they note a sudden increase in sales of model airplane glue. And teachers are told to suspect glue-sniffing if a child is irritable or inattentive or if he falls asleep or loses consciousness at school."[47] The witch hunt was officially on.

Many parents couldn't afford to take their children to the doctor every time they suspected something suspicious, especially if Glaser and Massengale's profile of low-income users was correct. Most store owners had learned since they put out their shingle that increases in sales of any individual product were fundamentally beneficial. And students had been falling asleep in class since the beginning of time. But now such actions would have to be viewed with a jaundiced eye. Every deviation from the norm could be a potential threat. Massachusetts was far away from the epidemic's ground zero out West, but even more troubling than the practice's obvious spread across the country was the fact that such fearmongering in a magazine with a circulation in the millions only ensured that the hysteria over the potential misuse of model airplane glue would continue to grow. Far more parents read *Consumer Reports* than they did the *Journal of the American Medical Association*. And with the hysteria that fed from such reports would come easy explanations of how to get high, bubbling over into more cases to be hysterical about.

The epidemic created the witch hunt, and the witch hunt only furthered the epidemic. It was, it seemed, an impossible paradox. The seemingly random discovery of children sniffing model airplane glue in Pueblo and other western towns had mushroomed in the few years since its discovery into a legitimate nationwide phenomenon. And like the mountains of Denver, it reminded everyone in its shadow of the frailty and vulnerability of those below.

# The Infection Spreads

MICHAEL J. WHELTON WAS frustrated. He read accounts of the growing menace of glue sniffing and worried that they all described the problem as endemic to the American West and Midwest. In November 1962, the physiology professor at Ireland's University College Cork wrote a letter to the editor of the *British Medical Journal*, reminding readers that sniffing glue was "well established all down the east coast of the U.S.A." He had worked at Johns Hopkins Hospital in Baltimore and saw many glue-sniffing cases parade through the hospital halls. The children had formerly found their highs by drinking codeine-enriched cough syrup but moved to glue once such products were restricted to prescription patients. Glue was readily available, and the peer pressure that developed among adolescents ensured that most were giving it a try. The "wave of this addiction" threatened to spread farther.[1]

By November 1962, however, the wave seemed to be everywhere. Throughout September and early October 1961, for example, Long Island police arrested more than a dozen New York kids found to be in a stupor caused by huffing model airplane glue. The Nassau County Health Department publicly warned that the active intoxicant in the glue was toluene. It was in every brand of model airplane glue, available in hobby shops and hardware stores everywhere. It could give you headaches. It could nauseate you. It could lead to liver or bone marrow damage. Irregular heartbeats. Coma. Fatalities had yet to come from such use, and Nassau County officials didn't know what constituted a fatal dose, but fatalities were sure to come.[2]

Still, the article claimed that glue could get users high, that it could provide "a feeling of elation similar to that of narcotics." Just as in Denver, the Nassau County Health Department's statement, published in the *New York Times*, could have provided a shot across the bow for study of a growing problem or it could have encouraged more use by clueing in youngsters that they had narcotic-grade substances in their desk drawers. Dire coverage of the phenomenon was the germ that fed the infection. Regardless, the rest of the New York media picked up on the story and began their own public declamations about the dangers waiting to destroy the youth of America.[3]

The same month, an eleven-year-old Washington, DC, boy went out into the woods to sniff glue, where he bumped his head and fell into a creek. His friends had to save him. But then again, his friends were largely responsible. The disease spread from youngster to youngster until it affected everyone. Soon, the boy began skipping school to sniff glue. His mother sent him to the doctor, then to the home of faraway relatives, but nothing seemed to stop him. "Once your child begins it," she told the *Washington Post*, "you don't know whether to trust him. You give him money for the movies, but you don't know how it will be spent."[4]

As of early January 1962, it wasn't just worried mothers sounding the alarm. Laurence L. Frost, director of the Washington Juvenile Court's child guidance clinic, had worked with several boys who had become problem glue sniffers. The product wouldn't addict a child like cocaine or morphine, but it was, Frost argued, still habit forming. He likened the personalities of juvenile glue offenders to those of adult alcoholics. "There is the same history of dependency—persons who cannot face up to things." Getting off of glue was far easier than similar attempts for harder drugs, but by the time some parents noticed the problem, permanent damage could already have occurred.[5]

More important than individual habits was glue's ability to move through a peer group, a school, a neighborhood. "It's sporadic," said Leo M. Allman, supervising director of Washington's junior and senior high schools. "It breaks out in a neighborhood, then it dies down. Every school has at least one or two who have tried it. We jump on it when we hear of it, but the stuff is readily available." Juvenile officials in Washington stated in early 1962 that they had seen well over a hundred reported cases in the preceding few years, dating to well before August 1959. Almost all of the cases involved children between the ages of eleven and fifteen, and the number of actual users must have been vastly higher than the number of reported cases. "No one seems to know how it came to Washington," the *Post* reported. "The Food and Drug Administration said it was first recorded about four years ago in

Los Angeles. The Public Health Service said it has had a few dozen reports from throughout the country." The FDA, for their part, began working with glue manufacturers in late 1961 to place warning labels on containers, but the labels were designed to prevent accidental exposure, not intentional inhalation, fitting the instructions proscribed by the Federal Hazardous Substances Act.[6]

The largest of those manufacturers was the Testor Corporation, which had its origin in the late 1920s when a Swedish immigrant to Rockford, Illinois, developed a nitrocellulose-based chemical for repairing shoes and women's stockings. Axel Karlson was hopeful that his product would be a success, but he knew he needed help, so he convinced the manager of the Woolworth's in Rockford's Swedish district, Nils F. Testor, to help him run the company. The original formula failed and Karlson moved back to Sweden, but Testor acquired the company's assets and founded the Testor Chemical Company. He retooled the chemical formula of Karlson's original adhesive, put it into tubes, and sold it as a bonding agent that applied to any number of household items. During the 1930s Testor began marketing specifically to hobbyists and model-makers. The company was a founding member of the Hobby Industry Association of America (HIAA) in 1940, and over the next two decades, it added model paints and models themselves to further corner the hobby market. In 1955 Testor acquired leading model manufacturer Duro-Matic Products and continued growing, to become the largest and most influential name in the model and model glue industry.[7]

When the epidemic began to spread—or when the narcotic use of model glue became a public concern—Testor had to respond. The company's president, Charles D. Miller, who had headed Duro-Matic before making the switch after the merger, assured the *Washington Post* that Testor's chemists were working to develop an adequate substitute for toluene, the dangerous aromatic hydrocarbon seeming to cause all of these problems. "We also plan to make the glue with an unpleasant smell," he promised, "but it is going to be more expensive to manufacture. We want to get our competitors to agree to the change at the same time."[8] That was the thing. If corporations were going to take the financial hit to develop a safer, less intoxicating glue, they wanted to make sure they weren't paying more than their competitors. Corporate entities were all in favor of competition in an unfettered free market capitalist system until their products proved dangerous and changes had to be made. At that point, cooperation seemed only fair.

Meanwhile, the locus of concern was still centered squarely on the children experimenting with glue. Soon after the *Washington Post* exposé appeared, a national audience first read about the pending problem. "Each year thousands of misguided teenagers explore the fuzzy-edged world of the cheap kick," *Time* magazine ominously reported. Cheap kicks had come in various forms over the years, but glue sniffing was the new problem. "You take a tube of plastic glue," the magazine quoted one fourteen-year-old addict as saying, "the kind squares use to make model airplanes, and you squeeze it all out in a handkerchief, see. Then you roll up the handkerchief into a sort of tube, put the end in your mouth and breathe through it. It's simple and it's cheap. It's quick, too. Man!" But not everyone was so pleased with the habit. "I don't like it," said one Salt Lake City teenager, "but I go back to it. If I could get liquor, I would. But it's too expensive and we can't get it anyway."9

Salt Lake City was another urban area whose "epidemic" had publicly exploded. Glue-sniffing arrests in Utah had risen precipitously over the first two years of the new decade. Twelve teenaged users were arrested and interviewed in the state capital in 1962, admitting to authorities that, though they had begun getting high by sniffing other substances (gasoline, nasal inhalers, ether), model airplane glue had become their drug of choice. It was a habit they considered to be extremely common among boys their age, though officials had yet to find any corresponding number of girls involved. Salt Lake City police argued that the teenage mania for glue was the cause of a spike in violence in the city. There were beatings, bottle fights. One glue-addled teenager tried quixotically to attack four marines. Such was an immediate danger, but the long-term effects were clearly there. The Poison Center of Salt Lake County General Hospital described significant kidney damage from overuse—and overuse was virtually guaranteed, because the more someone sniffed model glue, the more glue it took to get high.10

Of course, by "more glue," officials really meant "more toluene" or more "acetone" or more "butyl acetate"—more of the chemicals in the glue. Which meant that Testor was going to have to respond yet again. "We are going to change the formula," Charles Miller said, this time for a national audience, "by reducing the amount of acetone so that the narcotic effect will be slowed down, but I am afraid the kids will just switch to another product."11 The more jaded may have interpreted Miller's fear as a version of the competition versus cooperation conundrum, but it was clear that he was laying the groundwork for a defense against charges that his product was destroying America's youth.

The HIAA spent $250,000 in 1962 to combat glue sniffing. That year, the group produced a fifteen-minute film called "The Scent of Danger," describing the dangers posed by sniffing glue, cleaning fluid, nail polish, and other available solvents and recommending criminalizing the practice. It was a

Toluene is the most frequently used solvent in plastic cement or airplane glue, but the following volatile organic solvents may be used alone or in various combinations as a solvent and can act as an intoxicant:

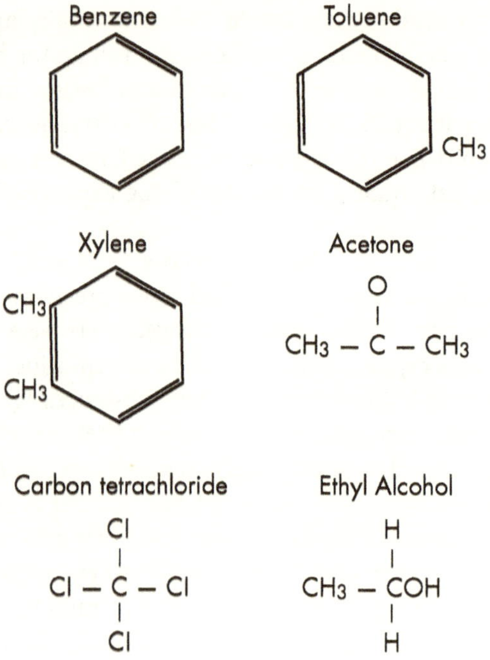

**Benzene**

**Toluene** — $CH_3$

**Xylene** — $CH_3$, $CH_3$

**Acetone** — $CH_3 - \underset{\underset{O}{\|}}{C} - CH_3$

**Carbon tetrachloride** — $Cl - \underset{\underset{Cl}{|}}{\overset{\overset{Cl}{|}}{C}} - Cl$

**Ethyl Alcohol** — $CH_3 - \underset{\underset{H}{|}}{\overset{\overset{H}{|}}{C}}OH$

toluene (toluol)
xylene
benzene
gasoline
naphtha
kerosene
methyl isobutylketone
methylethyl ketone
acetone
methylpropylketone
ethylenedichloride
trichloroethylene

methylene chloride
carbon tetrachloride
chloroform
amyl acetate
butyl acetate
ethyl acetate
diethylether
isopropyl alcohol
ethyl alcohol
butyl alcohol
hexane

VOLATILE SOLVENTS: This figure listed and diagrammed the most common solvents used in model airplane glue. Source: James L. Chapel and Daniel W. Taylor, "Glue Sniffing," *Missouri Medicine* 65 (April 1965): 288.

perfect combination for the hobby industry. Adults lauded glue manufacturers as responsible and concerned, while children who didn't read *Time* magazine learned how to get high from their model kits. Sales of model glue would rise steadily throughout the 1960s.[12]

This isn't to say that glue manufacturers were completely disingenuous. The association created a Glue Sniffing Committee chaired by Lewis H. Glaser, president of Revell, Inc., a Testor competitor. Working in cooperation with the Public Health Service, the committee began an effort to develop an irritant that would "make glue sniffing distasteful to young people." It was a vague claim, and it seemed in contradiction to Testor's goal of creating a bonding agent that didn't require toluene and other similar chemicals—but it was, if nothing else, a start.[13]

Meanwhile, *Newsweek* followed closely on *Time*'s heels, with its own exposé in August. "You're in outer space. You're Superman. You're floating in air, seeing double, riding next to God. It's Kicksville. Are these the fantasies of narcotic addicts on a pop? No. More disturbingly, these hopped-up reactions are those of teenagers hooked on goofballs, model airplane glue, and cough medicine." Newark, New Jersey, had suffered two teenage deaths from amphetamine overdoses. A twelve-year-old Miami boy was caught sniffing glue by his father, at which point the boy pulled a knife and threatened to kill him. The growth in media coverage, particularly national media coverage, gave the impression that the problem continued to grow. That more magazines and news outlets picked up the glue-sniffing story did not necessarily correspond with more children experimenting with glue, but this was certainly the impression that was conveyed. Robert Cooper, leader of New York City's Street Club Project, estimated that "of the 16,000 way-out adolescents we come in contact with, I'd guess that 15 per cent were glue-sniffing. It's on the increase."[14]

Prior to the new glue fad, kids had sniffed gasoline, they had "chewed on the inside of Benzedrine inhalers, and tried the Coke-with-aspirin routine," but glue was more dangerous. The liver, the gall bladder, and the bone marrow were at risk. Damaged nose membranes, blindness, even death could occur. "But this threat doesn't deter youngsters from squeezing up to five tubes of glue daily into a paper bag and breathing the fumes." It was easy to buy, obviously, but perhaps an even greater problem was the peer pressure on children and teenagers to avoid being "square."[15] The cool kids sought out such highs, and every kid wanted to be cool. This kind of domino theory of adolescent experimentation seemed undeniably valid, an easy explanation for the rapid spread of the practice, but again news outlets like *Newsweek* never really considered that their lurid coverage of "riding next to God" could ultimately act as the finger that tipped the first tile.

## Products Involved in Solvent Sniffing and the Chief Toxic Constituents of These Products*

*Plastic (styrene) Cements*
  Toluene †
  Acetone
  (Benzene)
  Aliphatic acetates (ethyl acetates, methylcellosolve acetates, etc.)
  Hexane
  Cyclo hexane

*Model Cements*
  Acetone †
  Toluene †
  Naphtha (petrolcuto oricin)

*Household Cements*
  Toluene †
  Acetone †
  Isopropanol
  Methyl ethyl ketone
  Methyl isobutyl ketone

*Fingernail Polish Remover*
  Acetone
  Aliphatic acetates †
  Benzene
  Alcohol

*Lacquer Thinners*
  Toluene
  Aliphatic acetates
  Methyl, ethyl or propyl Alcohol

*Lighter Fluid Cleaning Fluid*
  Naphtha (petroleum origin)
  Perchlorethylene
  Trichlorethane
  Carbon tetrachloride

*Gasoline*

---

\* Non-volatile and Substantially non-intoxicating constituents not listed.

† Principal intoxicants in various formulations

VOLATILE SOLVENTS BY PRODUCT: This list demonstrated the various toxic solvents in glues, cements, lacquer thinners, fingernail polish remover, lighter fluid, and gasoline. Source: Edward Press and Alan K. Done, "Solvent Sniffing: Physiologic Effects and Community Control Measures for Intoxication from the Intentional Inhalation of Organic Solvents, I," *Pediatrics* 39 (March 1967): 454.

The same month that the *Newsweek* article appeared, for example, a twenty-year-old man was admitted to London's Lambeth Hospital. He was an avid modeler who read about the glue-sniffing phenomenon in *Time* magazine and decided, based on *Time*'s coverage, to give it a try. He saw vivid colors, he began "seeing stories of space-fiction type in full colour," and began huffing his model airplane glue more and more. He soon found that when he tried to stop the practice, he would get abdomen pain, hand cramps, and other maladies. He was, for all practical intents and purposes, addicted. He finally reached the point where he was inhaling six tubes of glue per day, a pattern that ultimately led him to become semi-comatose, prompting his admission to Lambeth. After several hospital visits and a treatment of chlorpromazine hydrochloride, doctors were finally confident that they had broken him of the habit—a habit that began precisely because of *Time*'s coverage of the fad.[16]

Although the Lambeth case was exceptional, it scared city officials in urban areas that felt the brunt of the outbreak. The month after *Newsweek* published its exposé, in September 1962, George James, acting commissioner of New York City's Health Department, launched an investigation of glue sniffing. The study was sparked by city councilman Robert A. Low, who had called on the Health Department to ban the sale of model airplane glue to minors. James and his team inspected 1,746 hobby shops, hardware stores, and other variety shops in the metropolitan area, finding 52,265 bottles and tubes of model glue on the shelves. Even candy stores sold model glue.[17]

Availability was clearly a problem. In 1962 alone, glue-sniffing reports to the Youth Division of the New York City police department totaled 779. There were 443 more in the first three months of 1963. (By the end of 1963, that number would total 2,003.) Some stores were selling the glue to children along with paper bags for huffing it. Something had to be done. The Health Department finally acted in May 1963 and banned the sale of model airplane glue to anyone under eighteen, with the exception of tubes included in model airplane kits. Violation of the law would include a five-hundred-dollar fine or a year in prison. George James's study had indicated that the bulk of the problem centered in underprivileged areas such as Harlem, the Bedford-Stuyvesant section of Brooklyn, and the Lower East Side.[18] Still, every child was at risk. Model airplane glue was cheap and available, giving the lower classes access to cheap highs, but it was also part and parcel of the model-making hobby, a diversion centered on the middle and upper classes, those whose families could afford to buy model kits for their children. Harlem may have been hit harder than the Upper West Side, but all of New York's adolescents were at risk.

The city's attempt to legislate away the problem, however, was not the first. The initial anti-glue-sniffing law came from Anaheim, California, on June 6, 1962. Ordinance 1722 was very broad, making it illegal to "inhale, breathe, or drink any compound, liquid, chemical, or any substance known as glue, adhesive cement, mucilage, dope, or other material or substance or combination thereof, with the intention of becoming intoxicated, elated, dazed, paralyzed, irrational, or in any manner changing, distorting or disturbing the eyesight, thinking process, balance, or coordination of such person." The legislation didn't single out adolescents, it seemed to include drinking alcohol and any other form of intoxication, and it was vague as to punishment and scale of punishment for these various "crimes." A state law in Maryland coming soon on Anaheim's heels was more specific, singling out "any person under twenty-one years of age" for the intentional inhalation of substances "containing any ketones, aldehydes, organic acetates, ether, chlorinated hydrocarbons or any other substances containing solvents releasing toxic vapors." Any minor caught breaking the law would be charged with a misdemeanor and fined. The Maryland statute was far more specific as to punishment and the nature of the illegal substances, and it emphasized juveniles as being the principal abusers of such products. Still, the laws of Anaheim and Maryland were fundamentally different from the New York legislation. They punished users, while New York punished sellers. As more cities and states continued to work through their own fumbling attempts at legal remedies to a social problem, they would almost always divide along the same lines. And it should come as no surprise that the HIAA only supported those laws in the Anaheim and Maryland mold. For association members, it was improper use that was the real menace, not those who sold the product.[19]

Regulation of cocaine had happened much the same way earlier in the century. Newspaper reports fed public fears of a nationwide drug menace, which created a reciprocal relationship between them and activists who sought an end to various forms of drug abuse. They were joined by regulators in the fields of public health, social work, and child welfare who sought regulations that would help control such abuse. It was, according to Joseph Spillane, an effective coalition. But it did have one element that was absent in the attempt to regulate model airplane glue. The pharmaceutical industry, which had produced cocaine as an anesthetic, willingly divested themselves of their interest in cocaine and joined the coalition in an effort to influence legislation successfully in a manner that would give them an advantage over manufacturers of patent medicine. Hobby companies like Testor and the HIAA, on the other hand, worked to find a suitable astringent for their products but did not join the hue and cry for legislation to regulate the product. Unlike the early century pharmaceutical companies, hobby companies found no ancillary competitive benefit in supporting such attempts at legal regulation.[20]

Because of the involvement of the hobby industry, because of glue sniffing's clear tie to children, and because of the indecision of many as to whether model airplane glue was actually a drug, battles over such legislation could be controversial. Washington, DC, which felt the epidemic early, provides a good example. In November 1962, Detective Robert Wrenn of the District's Juvenile Bureau said that he had witnessed more than three hundred cases of glue sniffing in the previous two months alone. In December, three teenage boys and a teenage girl ran away from home, got high on model glue, then went on a car theft rampage. They stole a car and drove it until they crashed, then stole another, until all together they had stolen fifteen cars. In the fifteenth, they led police on a long chase before crashing into a telephone pole. The girl—the first such girl to be publicly discussed as a glue sniffer—tried to jump from a third-floor window after being taken to police headquarters.[21]

Things were getting out of hand. In January 1963, the District's Board of Commissioners began discussing punitive measures. Basing their own ordinance on the earlier Maryland law, they proposed to prohibit sale to minors without written consent from parents, to require merchants to keep records of glue purchases and purchasers, and to make merchants contact police when anyone bought more than two tubes of glue at one time. There would also, like the Maryland law, be a provision to make the act of sniffing glue itself illegal. Deputy Police Chief John E. Winters supported the legislation, arguing (against the assumptions of New York officials) that glue sniffing was a decidedly middle- and upper-class problem. And the kids were being caught everywhere—at recess during school, at the bowling alley, in the park. They were committing robberies and assaults under the influence. A citywide regulation was the only way to curb such behavior.[22]

Police officers took to the radio to discuss the dangers of sniffing glue. Newspapers reprinted reports from around the country and the world of dangerous glue-sniffing debacles. Winters issued a report responding to concerned parents, noting that police had been called to Washington public schools 270 times from the start of the school term in September until the second week of February 1963. Still, progress was slow. No legislation appeared in the first half of 1963. Finally, in June, corporation counsel Chester H. Gray completed a draft of the regulation, and the commissioners set a public hearing for July 19.[23]

It turned out that not everyone was so happy about the proposed bill. The HIAA had been consistent in selling to the public its own attempts at self-policing the problems stemming from model glue, but it had no use for Washington's own police attempt. Lee L. Blyer represented the HIAA at the hearing. He argued that model glue was just one of many household products that could be huffed for intoxication. Singling out model glue would give hobby shops "the stigma of merchandizing products injurious to health." Blyer argued for regulations based on the Anaheim ordinance, prohibiting sniffing but not punishing sellers. Besides, he argued, if the glue manufacturers were successful in creating a less toxic product, all of this regulatory action would be unnecessary. Philip Corr, a local hobby shop owner, agreed with Blyer. "We don't want parents to think that because their youngsters are building models they are going to become degenerates."[24] It may have been a sentiment that came from blatant self-interest, but it was certainly a fair point.

But hobby shops weren't the only organizations that felt their interests challenged by the District's proposed legislation. Later in July, the US Navy took its own stand against the bill. Commander Paul Boyer, the Pensacola Naval Air Training Command public information officer, argued that limiting children's access to models "cannot help being indirectly deleterious to the already hard-sell aviation and aerospace recruiting programs of the armed forces." Model airplanes spawned the dreams of young would-be pilots. Model airplanes made kids want to fly. "Here they learn flight fundamentals and acquire the ambition and desires to be pilots of tomorrow."[25] It seemed that the navy's argument was a reach at best, but the power of such charges was undeniably powerful. No public official was interested in publicly weakening the military.

Still, despite the protests of the military, it was clear that something had to be done. Juvenile arrests jumped 13 percent in 1963, and the Metropolitan Police Youth Aid Division's annual report laid much of the blame for truancy, burglary, and other more violent criminal offenses at the feet of model airplane glue. Just weeks after the August report, a sixteen-year-old Prince George's County boy died after inhaling model glue from a plastic bag. He passed out and suffocated with the bag over his head. The crime statistics and the death prompted city commissioners to move toward some kind of resolution, and the overwhelming evidence in front of them lent weight to the idea that restrictions on sales were just as necessary as punishment for sniffing itself. Walter Tobriner, commission president, made the point that few children would know the resolution and the legal consequences of their behavior. Adults were the responsible parties, so adults needed to be the principal locus of regulation.[26]

Finally, on February 18, 1964, the Washington, DC, commissioners passed the regulation. It followed the original plan, that is, to make glue sniffing illegal, but it also restricted sales to minors without written parental consent and required records of sales of more than two tubes of model glue. The regulation also banned street sales of model glue. Violations could result in three-hundred-dollar fines and ten days in jail.[27] The final product was almost exactly like the original proposition of January 1963, but controversy had delayed passage for more than a year. The real question at the center of the peripheral debates about the thrust of the legislation was whether glue was really a drug. The hobby companies, the US Navy, and shop owners argued that it wasn't—that there was a viable, necessary use for the product they sold, and that they should not be punished because a teenager decided to use that product improperly. But the criminal activity, the truancy, the deaths, all seemed to argue differently. Such were the creatures of hysteria, and epidemics feed on hysteria. Still, Washington's hysteria functioned differently than that of its fellows. Law enforcement coordinated that hysteria for the express purpose of getting its regulation. That kind of instigation from a moral agent is more reminiscent of the marijuana push of the 1930s, but whether the panic was coordinated or developed from a growing, nebulous parental concern in Washington, and in many other cities and states following it, panic would always win.

In January 1963, as Washington's officials first proposed to legislate its glue menace away, *Consumer Reports* produced its own exposé on the problem, its own contribution to the panic (see chapter 2). "A novice sniffer may find he produces a full-blown 'jag' with the fumes of one tube of glue, but a physical tolerance builds up with repeated use so that it may eventually take as many as five tubes to produce an effect." The *Consumer Reports* article quoted heavily from the Denver Glaser and Massengale study, but that was an important task. Far more people read *Consumer Reports* than read the *JAMA*, and the notion that the respiratory tract, nervous system, heart, liver, kidneys, and bone marrow could all be affected by the practice of sniffing glue was only spread wider through the popular magazine. The practice also caused "unpleasant breath odor," the magazine reported, "and excessive secretions from irritated mucous membranes of the nose and throat may lead to frequent spitting."[28]

With a national witch hunt now in full flower, New York Mayor Robert F. Wagner petitioned the President's Advisory Commission on Narcotics to offer federal grants to urban areas in order to supplement their spending on narcotics enforcement. Heroin and other opiates were the most pressing and expensive problems faced by cities, but Wagner pointed to model airplane glue as a dangerous precursor to such harder materials, only increasing the likelihood that children and teenagers would graduate to more dangerous substances and the inevitable criminal activity that stemmed from such use.[29]

But New York was quickly discovering that model glue could, at times, be just as dangerous as the harder drugs. In December 1963, a fourteen-year-old Brooklyn boy stole away to the roof of a local factory to sniff glue, but in his intoxicated state, he lost his balance and fell seven stories to his death. Another similar death happened a few days later. The following week, the New York Board of Health, the policy-making arm of the city's Health Department, approved an amendment to the city health code to ban glue sales to minors, building substantively off the proposed Washington, DC, regulation (which wouldn't be passed until February). But the New York amendment would go farther. Hobby clubs of more than twenty-five members would have to register with the Health Department in order to purchase glue. Punishment would include fines as high as five hundred dollars and a year in prison. The board announced that it would happily take suggestions from the hobby lobby and the chemical industry, but something was going to have to pass. In 1963 alone, New York officials documented more than 1,800 glue-sniffing cases, up from 800 in 1962.[30]

Ultimately, there was nothing the HIAA or glue manufacturers could do. Hysteria over the deaths of children sniffing glue had kept the dominos falling farther from them at every turn. Here again city officials directed the flow of information, and the media served as a conduit to frighten local parents. Of course, those officials were parents too, and the statistics were legitimately frightening, which only made the panic more powerful. The health code amendment passed easily. In late February, Michael Vance, a Bronx teenager, got high sniffing glue and then tried to jump the ten-foot distance between the roof of his apartment building and the neighboring structure. He didn't make it. Vance was the third teenager in as many months to fall from a building after sniffing glue. The police, now armed with the health code amendment, charged Joseph Kouri, a candy store owner who sold Vance the glue, with "impairing the morals of a minor" and violating the city health code.[31]

So the epidemic spread beyond its western bounds, invading urban populations across the country. By 1963 glue sniffing was on the minds and tongues of parents and officials in virtually every American region. Cities and states began experimenting with legal actions to stop the problem but had difficulty generating the kind of mass support needed for such action, because ultimately they wanted to criminalize a product designed to be in the hands of children. The corporations with direct ties to model airplanes had their own agendas, clearly wanting adolescent misuse of their products to stop but still wary of measures that put the burden on the merchants. It was clear as of 1963 that the country was not dealing with a passing fad. Inhalation of model airplane glue had only escalated since the warm summer of 1959, and as more news outlets reported on the problem, more adolescents discovered the possibilities and decided to give it a try. Here again was the great paradox. The problem could only be solved through discussion, but discussion only put glue sniffing into more American homes. Such conundrums were the stuff of science and social science, and while the popular media issued its dire warnings, doctors and academics struggled with the contours of a phenomenon that existed well beyond the doors of their hospitals and labs.

# The Academy Responds

INFECTIONS SPREAD THROUGH HUMAN networks based largely on opportunity, proximity, and vulnerability. Whether tuberculosis or gossip, contagions diffuse in surprisingly predictable ways. Mary, for example, has been sleeping around. Elizabeth, who has always been jealous of Mary, begins calling her a whore. Of the ten people in Elizabeth's inner circle, three are friends of Mary and tend to keep to themselves. Three like Mary but also love to share information with their friends. The other four never liked Mary to begin with. The contagion will die with three of Elizabeth's friends but spread with the seven others. When the friends move outside of Elizabeth's circle, they will spread the contagion to their own social groups, where there will be comparable ratios of spread and cessation. Soon, Mary will be a whore all over town.

The difference between tuberculosis and Mary being a whore, of course, is that the opportunity, proximity, and vulnerability to tuberculosis are much greater than they are to Mary's recreational choices. Assaults on Mary's reputation will end, for the most part, with people who know Mary (or, in the modern world, with people who are Facebook friends with people who know Mary). Vulnerability ends at that point. But tuberculosis can continue spreading beyond the borders of those familiar with the disease. The Centers for Disease Control performs a procedure known as contact tracing to manage such outbreaks. The virology can be mapped (see facing page), with the black dots representing infected and infectious persons, lighter gray dots representing dormant infections, and the darker gray dots representing potential victims.[1]

CONTACT TRACE MAP

On any scale of potential spread, "sniffing glue" would map to a point far closer to "tuberculosis" than "Mary is a whore." Like gossip, sniffing glue required a willingness to engage in the activity, and vulnerability to the disease seemed limited to a certain segment of the population (namely, adolescent males), but opportunity was everywhere, meaning that any contact tracing map could spread to every corner of the country, to every corner of the globe.

Any such map would require more than scare headlines and fantastical coverage to make predictive analyses of the contagion's spread. Opportunity, proximity, and vulnerability required analysis, study, and an accurate portrait of potential victims. To this end, researchers Gordon Barker and W. Thomas Adams returned to the presumed genesis point of the epidemic, Colorado. They studied glue-sniffing and non-glue-sniffing internees at the Lookout Mountain School for Boys in Golden, Colorado, and sought to discover statistically significant tendencies among glue sniffers, to find the potential gray dots on the map. (They also checked the Mountview School for Girls in Morrison, Colorado, but found none who sniffed glue. Despite scattered accounts of female glue sniffing, the problem from the outset seemed decidedly male.)[2]

All of the boys came from homes with pronounced "family disintegration"—single-parent households dependent upon one form of welfare or another and with siblings or other members of the home with criminal trouble.[3] Even when both parents were in the home, domineering fathers with low-paying unskilled labor jobs ensured some form of instability. Also, the glue sniffers tended to have far more siblings, residing in homes with an average of 7.5 children, as opposed to 4.5 children for the control group. Almost half of the glue sniffers were thirteen years old or younger. All but one came from either Denver or Pueblo Counties, significant urban centers of the state. While some of the control group came from these areas, others came from rural areas (in a predominantly rural state). The researchers explained the urban nature of their findings by hypothesizing a "lack of access to the materials, stronger social controls, or lack of knowledge of this particular behavior" among rural boys. All this translates to opportunity, proximity, vulnerability.[4]

Glue sniffers were also poor students and "showed far greater retardation educationally than the control group." Based on California Achievement Test (CAT) scores, the glue-sniffing group was at least two grades behind the control group. "Most of the boys who used the glue were practically non-readers with only primitive academic skills." They were truants. They were dropouts.[5]

They were also, the report suggested, Latino. The vast majority of the glue sniffers under examination came from a "Spanish-American" heritage, "thus it is fairly clear that detected cases of glue-sniffing in Colorado are found predominantly in one ethnic minority."[6] Demonstrating family

Siblings in Families of Glue-Sniffers and Non Glue-Sniffers

| No. | 0 | 1 | 2 | 3 | 4 | 5 | 6 | 7 | 8 | 9 | 10 | 11 | 12 | 13 | 14 |
|---|---|---|---|---|---|---|---|---|---|---|---|---|---|---|---|
| Glue Sniffers | 0 | 0 | 2 | 0 | 2 | 6 | 3 | 2 | 6 | 1 | 2 | 0 | 2 | 0 | 2 |
| Control Group | 0 | 1 | 2 | 3 | 8 | 7 | 2 | 2 | 1 | 1 | 0 | 1 | 0 | 0 | 0 |

D = 11, p < .05. See Sidney Siegel, *Non-Parametric Statistics for the Behavioral Sciences* (New York, McGraw-Hill, 1956), 278.

GLUE SNIFFING AS A FUNCTION OF THE NUMBER OF SIBLINGS IN A GIVEN FAMILY: Barker and Adams demonstrated that glue sniffers tended to have far more siblings than non-sniffers. Source: Gordon H. Barker and W. Thomas Adams, "Glue Sniffers," *Sociology and Social Research* 47 (April 1963): 301.

### Grade Placement of Glue-Sniffers Compared to Non Glue-Sniffers

| Grade Placement | 10+ | 9.5–8.6 | 8.5–7.6 | 7.5–6.6 | 6.5–5.6 | 5.5–4.6 | 4.5–3.6 | 3.5–2.6 | 2.5–2.0 |
|---|---|---|---|---|---|---|---|---|---|
| Glue Sniffers | 0 | 1 | 0 | 1 | 3 | 11 | 3 | 5 | 4 |
| Control Group | 1 | 3 | 7 | 6 | 5 | 3 | 1 | 2 | 0 |

D = 11, p < .01. Op. cit. p. 278.

GLUE SNIFFING AS A DETERRENT TO ADEQUATE SCHOOLWORK: Barker and Adams demonstrated that glue sniffers tended to be at least two grades behind the non-sniffing control group. Citation references Sidney Siegel's *Nonparametric Statistics for the Behavioral Sciences* (New York: McGraw-Hill, 1956). Source: Gordon H. Barker and W. Thomas Adams, "Glue Sniffers," *Sociology and Social Research* 47 (April 1963): 304.

deterioration or low CAT scores provided useful variables for determining the makeup of the average victim of the epidemic, but assuming that Latinos were more vulnerable than other ethnic groups did not seem to match other accounts. Barker and Adams quoted Jacob Sokol, chief physician of Los Angeles County Probation's Juvenile Hall, that "glue sniffing among the Negro population as compared to the Caucasian and Spanish-speaking population is very small." Their conclusion, then, was that cultural factors relating specifically to Colorado created the ethnic nature of the state's epidemic, not an inherent genetic draw. Still, the study ruled out middle-class boys as chronic users. "Glue-sniffing parties" among middle-class boys received consistent media attention, but they were "short-lived and do not occur at regular intervals with the same participants." It was analysis that seemed shortsighted at best, bigoted at worst.[7]

Perhaps a better control group would be juvenile drinkers. "Drinking of alcoholic beverages among youth," noted James W. Sterling, an official with the Chicago Police Department's Youth Division who produced his own portrait of the potentially infected, "is so pervasive in contemporary American society that it can be considered a normative mode of achieving intoxication." To that end, Sterling studied forty-seven glue sniffers and fifty adolescent drinkers on the initial assumption "that there are no statistically significant group differences between juvenile glue sniffers and juvenile drinkers as measured by selected personal and social background factors."[8]

Sterling's hypothesis, however, proved to be slightly off. The mean age for Chicago sniffers was twelve years, three months. The mean age for juvenile drinkers was thirteen years, eleven months, which reflected "the first evidence of deviance on the part of the sniffers, to wit, that they first became involved with the police about 1½ years earlier than the drinkers." Sterling also noticed the tendency for sniffers to be male, as only seven of forty-seven glue sniffers were female (a much higher total than in other studies) even though females committed 20 percent of the juvenile offenses in Chicago. In addition, there were no black offenders in the Chicago ranks, a surprise considering the arrest ratios of African American drug users. Sterling was at a loss to explain it. "Is this the result of a differential in police concern over youthful intoxicated behavior in White and Negro areas?" he asked. "Do Negro juveniles have easier access to beer, liquor, and wine than Whites? Is sniffing an activity which affords the Negro little or no esteem from his peers, and hence its practice is not consensually supported?" They were good questions, though the first seemed most likely. Black youths *did* sniff glue, after all (see chapter 8). In addition, like Barker and Adams, Sterling also found a propensity toward glue sniffing among Latinos.[9]

A clearer picture was taking shape. Glue sniffers seemed to have a relatively unique profile. Still, the Chicago study found no significant difference between juvenile glue sniffers and drinkers when it came to family background (commonly assumed to be a problem well within the purview of sniffing). Both groups also showed similar rates of recidivism, and both groups tended to enjoy their drugs of choice in groups. Both, too, showed maladjustment at school, though alcohol users tended to drop out while glue sniffers tended to demonstrate "retardation in grade level," a phenomenon demonstrated by the Colorado study. There was no significant difference in the kinds of ancillary crime that both activities encouraged. In fact, the only other real difference between the two groups and thus the only other etiological factor in the glue-sniffing portrait was that drinkers tended to exist in relatively equal numbers across the city. Glue sniffing was limited to certain neighborhoods in what Sterling called "clusters." This was not to say that the practice couldn't spread to any region of the city. But sniffing tended to catch on in certain groups, obsessing kids in one neighborhood before moving on to another.[10]

But how exactly did it move on? Why did the practice spread? What was the contagion? Sterling had no answers. Barker and Adams could only speculate. Glue could be a forerunner to alcohol, cheaper and more accessible for those too young to buy beer. It could be an escape from troubling family situations. Glue sniffing could garner attention for children drowned out by an excessive number of siblings. Social pressures could then take over to spread the practice through the population.[11]

There was no way to prove such theories. But in every epidemiological study, no matter the disease, just as important as understanding the profile of potential victims is understanding the effects of the disease on those who already have it. Cries of potential long-term neurological damage had been shouted from the rooftops. Glaser and Massengale, along with others, had speculated about permanent brain, kidney, or liver damage. Rex Wilson had reported rare (but existent) cases of aplastic anemia. Maybe there was electroencephalographic change. Maybe there was cerebellar degeneration from chronic toluene sniffing.[12] But though worried declamations about the possibilities of such long-term effects seemed to gain momentum because of the scare they gave to both scientists and parents, little evidence of actual permanent damage existed. To that end, Josiah Dodds and Sebastiano Santostefano, researchers with the University of Colorado Medical Center, sought to see if there was any real weight to the claims.

There wasn't. Dodds and Santostefano studied a group of glue sniffers from Denver's Juvenile Hall and a control group of randomly selected Denver schoolchildren. They tested attention, concentration, perception. They tested memory. They tested performance in the wake of distraction. They tested visual motor coordination. They tested spatial relationships. "It seemed reasonable to believe that any differences observed between the two groups might be related to the effects of glue-sniffing on cognitive functioning," but there was no significant difference. On motor coordination. On concentration. On spatial relationships. On anything. With that kind of comprehensive examination of cognitive skills and functions, it seemed clear to Dodds and Santostefano that glue sniffing did not cause brain damage.[13]

But how could such claims be justified when previous researchers were talking about long-term consequences? The researchers defended against that question by arguing that all of the sniffers in the examination were chronic users—they had ingested substantial amounts of glue for several years. Even if the long-term consequences of such use had yet to manifest themselves, it was clear that precursors to such consequences would already have shown themselves in some form. (The one hitch in the Dodds-Santostefano manifesto was that other researchers had seen plenty of evidence of distraction, tremors, deteriorated motor skills, and other problems. It is significant that the University of Colorado Medical Center study's glue-sniffing group was composed of only fourteen adolescents.) The other significant question that stemmed from such analyses was the possibility that sniffing glue affected parts of the central nervous system (the cerebellum, for example) not detectable with tests of concentration and motor skills. Again, however, the researchers noticed no tremors, no "peculiarities in gait." There was, to be sure, psychological difficulty that still required plenty of study, but there was, they argued, no permanent cognitive damage brought about by sniffing glue.[14]

But what about other problems? An editorial in the *Journal of School Health* suggested that glue sniffing could cause long-term blood disorders. Aplastic anemia seemed to be the most persistent problem, limiting the ability of bone marrow to produce cells in sufficient number to replenish blood cells. Unlike a general anemia, which typically limited the body's production of red blood cells, aplastic anemia ultimately limited the production of both red and white blood cells, as well as blood platelets. Darleen Powars, a pediatric hematologist at Los Angeles County General Hospital, studied five cases of adolescent glue sniffers, all of whom demonstrated "an erythrocytic aplastic crisis" marked by anemia, reticulocytopenia, and hyperplastic

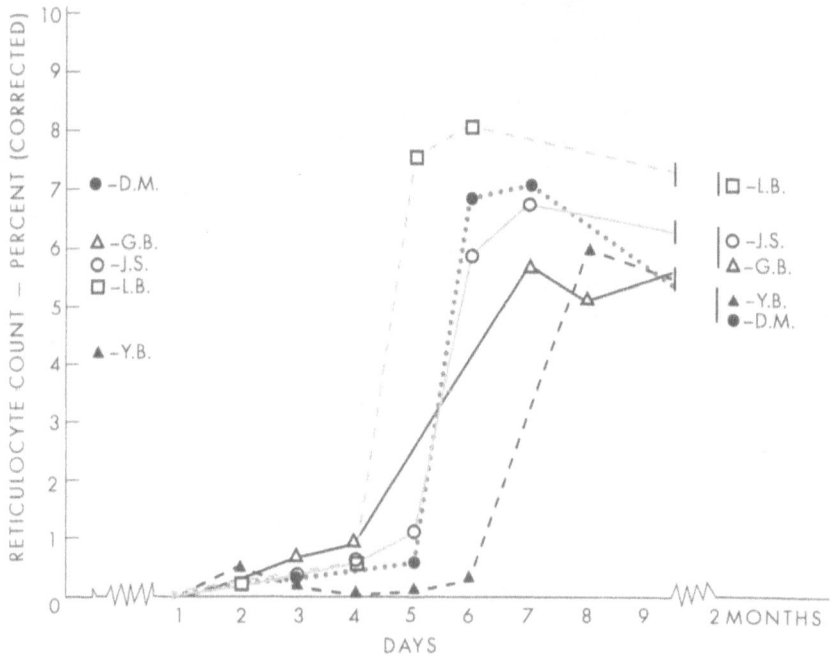

Aplastic Crisis after Glue Sniffing in 5 Patients with Sickle-Cell Anemia (Reticulocyte Response)

APLASTIC ANEMIA AMONG GLUE SNIFFERS: Darleen Powars demonstrated a massive jump in reticulocyte count in glue-sniffing patients with sickle-cell anemia. Source: Darleen Powars, "Aplastic Anemia Secondary to Glue Sniffing," *The New England Journal of Medicine* 273 (September 23, 1965): 701.

bone marrow. All of the patients were black, seemingly exploding the profile established by earlier studies that argued that glue sniffing was not an African American problem, but all also had sickle-cell anemia, a notorious problem among the black community that could cloud potential prognoses for long-term blood effects for noted glue sniffers.[15]

As early as 1941, however, researchers had demonstrated that benzene ingestion in rats had caused erythroid and myeloid hyperplasia of the bone marrow ultimately leading to death. Patient results of the Powars study seemed to validate such conclusions. Of course, ingestion of the amounts of toxic benzenes was a hit or miss proposition, but it seemed as though the "amount of benzene necessary for bone-marrow aplasia to develop" would require sustained use for at least six months. That was for regular patients. For patients with sickle-cell, damage to "the erythropoietic function of the bone marrow" was quickened, leading to aplastic crisis in a matter of days.[16] Jacob Sokol agreed. "There are changes in form, shape, and color of the red cells, increase in the number of white cells, and decrease in white blood count." Basophilic stipplings and target cells indicated severe toxicity. Bone marrow aspirations indicated an inhibition of the maturation of blood cells. On top of that, Sokol found liver and kidney damage (though he made no claims of long-term damage), and "urinalysis reveasls pus, albumin, casts, bacteria, and blood." The problem might not always cause severe internal damage, but it could.[17]

Powars's and Sokol's data appeared late in 1965, demonstrating significant progress over the first half of the decade in detailing the potential health risks associated with the spread of the epidemic. But in many ways, interpretation of the problem still stagnated, emphasizing the same concerns fretted over in the early part of the decade. Broken homes were strong indicators of potential use. White and Latino boys were far more likely to sniff glue than were African Americans or girls. There were, however, additions to the epidemiological portrait. "Some of the early symptoms to look for," argued Sokol, "might be day-dreaming problems of life, guilts and anxieties and worry about school and friends or marks and grades, bashfulness, self-consciousness, and other manifestations of inadequacies or inferiority." The frustrations over an inability to meet goals. A lack of love. The boys who chose to sniff glue were usually small of stature, they weren't athletic.[18]

Still, what seemed like progress in creating a fuller profile of the glue sniffer was complicated by the fact that academic work on the dilemma wasn't coalescing into a full consensus. Willie S. Ellison, of California's Santa Clara County Juvenile Probation Department, witnessed dramatically lower intelligence quotients among glue sniffers, for example. He saw weak personalities, boys under the age of fifteen (with a slow rise in the number of girls participating in the practice), criminal behavior, and high rates of recidivism. At the same time, however, his cases demonstrated sustained use among lower-class boys from minority groups.[19] If any attempt to accurately describe the black dots in a contact trace and keep them from spreading was going to be successful, ethnicity and socioeconomic status were going to be fairly important variables.

The economic benefits of the Second World War had largely been squandered by conservative economic policies in the 1950s. American unemployment hit a low of 2.9 percent in 1953 but rose steadily through the Eisenhower and Kennedy years. Lyndon Johnson would inherit an unemployment rate hovering near 6 percent but would reduce it every year. By 1969 American unemployment was at 3.5 percent (though the Nixon administration would set it on another rise almost immediately).[20]

Average weekly income continued to rise incrementally throughout the entire period from the end of World War II through the Nixon administration, but a diminution in unemployment and a rise in weekly income belie a steadily growing wealth disparity. A massive 1964 tax cut increased wealth at the top of the economic spectrum. Meanwhile, corporate taxes in 1973 were half of what they were in 1949, while the percentage of Social Security taken from American paychecks had more than doubled. Johnson declared a war on poverty, in an attempt to compensate for some of the more insidious problems related to the growing wealth gap, and it did have an effect. The number of people below the poverty line fell steadily throughout the 1960s and began to rise again only after Nixon's takeover. More important for any examination of the abuse of model airplane glue, the number of dependent children under eighteen below the poverty line also fell by more than 40 percent throughout the decade. There were broad racial disparities in such numbers, however. In 1967, for example, 11 percent of the white population was below the poverty line, compared to almost 40 percent of the black population (strong numbers on the Hispanic population don't really exist until the early 1970s).[21]

Those disparities existed for school enrollment and similar environmental factors for potential glue sniffers. The most likely glue sniffer was a male child aged between ten and thirteen. White students in that demographic were about 4 percent more likely to attend school than black or Hispanic students. Above the poverty line, they were 5 percent more likely to attend than their black counterparts and just less than 4 percent more likely to attend than their Hispanic counterparts. Below the poverty line, corresponding differences were 2.3 percent and 4 percent. Significantly, however, no group of ten- to thirteen-year-old males, factoring for poverty and race, attended school at a rate of less than 90 percent. Poverty and race could affect quality of attendance, enrollment, rates of delinquency, but they did not have a dramatic affect on enrollment. And that was significant. Schools are incubators of children's ideas, both good and bad. They serve as a grapevine, a facilitator of any related contact trace map.[22]

Glue sniffing, whatever the makeup of its adherents, was, for Ellison, "a symptom of a large social problem." It was "a form of passive retreat taking the individual away from the family, and any other organized form of community life." Ellison thought the best way to combat the problem was to "withdraw the habitual sniffer from psychological dependence by placing him in Juvenile Hall for about three weeks." But if sniffing was part of a larger social problem, more was going to be needed. Sokol sought a comprehensive program of arousing the attention of the public, government officials, school boards, and manufacturing companies. He wanted more sponsored research. He wanted probation officers and doctors to speak to youth about the potential dangers of glue sniffing. And he wanted lobbying efforts to convince the HIAA to "do its utmost" to remove toluene from model airplane glue. Powars, meanwhile, saw a way out through discovering just what products glue sniffers preferred. The product itself was, after all, the line that connected the black dots to the varying shades of gray. The three most common and most preferred benzene-ring-based model glues were Testor's Polystyrene Plastic Cement, Revell Cement, and LePage's Liquid Solder. The first two were loaded with toluene, the last with acetone.[23]

But any description of the spread of such substances would ultimately rest on whether such products were addictive. The movement of any disease through a population depended heavily on whether that disease could take hold. The World Health Organization argued that any chronic (or even periodic) intoxication from repeated use of a drug or intoxicating agent that proved detrimental to individual or societal health was an addiction. Overpowering desires to use such substances, a presumed dependence, and a tendency to continue increasing the dosage of those substances (generally caused by the inability of smaller doses to produce similar bouts of intoxication after prolonged use) all played a role. Such definitions, however, served the cause of public warning far more than they did scientific evidence. As early as the late 1930s, sociologist Alfred R. Lindesmith posited that wrongly equating withdrawal with addiction only fostered addictive behaviors. Addiction, in Lindesmith's conception, was a learned process over time, wherein patients connected withdrawal to drug use, then began engaging in that drug use specifically to avoid the withdrawal experience. Addiction

## Ingredients of Some Airplane Glues Available in California

*Testors Model Cement Formula A*
    Cellulose acetate                Acetone*
    Tricresyl phosphate*

*Testors Formula B*
    Cellulose nitrate                Acetone*
    Tricresyl phosphate*        Butyl acetate
    Hexane*                      Butyl alcohol*
    Isopropyl alcohol              Toluol*
    Isopropyl acetate

*Revel*
    80 percent toluene*

\* Indicates toxic product.

INGREDIENTS OF SOME AIRPLANE GLUES: Tricresyl phosphate was not only used in model glues, but had a long history of being an active ingredient in illegal liquors during the Prohibition era. Source: Martin L. Barman and Donn B. Beedle, "Acute and Chronic Effects of Glue Sniffing," *California Medicine* 100 (January 1964): 19.

wasn't the result of exposure to drugs like, for example, heroin. It was the result of a mental shift in a person's motivation for that exposure.[24]

Ellison found no evidence of what he termed addiction for model airplane glue, but he did note that people "can become psychologically dependent upon glue." According to Ellison and other American researchers, this wasn't the same thing. Leland M. Corliss, director of health services for Denver public schools, noted that tolerance to glue did build over time, and that there was a legitimate psychological dependency, but "true addiction in the sense of addiction to heroin has not been found." Corliss worried that terms like "addiction" were thrown around carelessly when referring to model airplane glue. "Some people," he claimed, "really get carried away with themselves." Similarly, Edward Press, editor of the *Journal of Pediatrics*, understood that tolerance to glue increased with repeated use and that many used it repeatedly. Such, however, seemed to be angels dancing on the heads of pins. Psychological dependency and a necessity to increase dosage to improve narcotic effect were (regardless of any specific labels medical professionals wanted to ascribe to them) clear addictive behaviors. Or, as Alfred Freedman and Ethel Wilson stated in an address to their New York Medical College Department of Psychiatry: "The most significant distinguishing characteristic of addiction is the qualification of *harm* to oneself or to others produced by the regular or habitual use of any substance. The production of physical dependence, with its concomitants of tolerance and withdrawal symptoms, is often, but not necessarily, present." Addictive behaviors encouraged the spread of the use of model glue, so there was really no point in calling its use anything other than addiction.[25]

Again researchers omitted the fine lines when discussing model airplane glue. Unlike, say, heroin (they seemed to be arguing), addiction wasn't a foregone conclusion with sniffing glue, but addictive behaviors were just as dangerous. "The emotionally well-balanced and well-adjusted child," argued Edward Press, "might experiment once or twice with glue sniffing as a result of a combination of curiosity, boredom, and availability, but he would not be likely to indulge in repeated bouts."[26] Maybe that was true and maybe it wasn't. But Press's larger point seemed unquestionably valid. While many kids sniffed glue out of peer pressure, curiosity, or any number of other reasons, the number that became glue sniffers in a larger, more consistent sense was far lower. Far from being encouraging, however, such discrepancies only posed more problems. With heroin, researchers assumed that the spread of infection was more easily predicted because just a few uses grew to many very quickly. Heroin in any one area allowed those who studied it to presuppose spread throughout that area. But if glue use only proved chronic (if not addictive) in only certain cases of childhood experimentation, then predictive models were far more difficult to draw.

It was, to be sure, a flawed comparison, one that only increased the problems with any potential predictive model. Lindesmith's studies had essentially debunked this comparative model of heroin addiction. The Vietnam experience would only further problematize it in the future, as researchers discovered that heroin users in Southeast Asia remained addicts upon their return from service only in extremely rare cases.[27] A similar addiction discussion was progressing for researchers dealing with users of amphetamines, who fell into a semi-comatose state when denied drugs, the opposite kind of withdrawal reaction that came from users of opiates or alcohol. Regardless, as Nicolas Rasmussen had demonstrated in his study of those amphetamines, as of the late 1960s the notion of addiction was "still a pivotal issue politically if not medically."[28] That's not to say there wasn't still controversy. Even into the 1970s, several American researchers dealing with substances ranging from alcohol to speed were still arguing that addiction was a singular disease that applied to all substances in more or less the same way.[29]

A slightly more effective comparative analysis argued that, while heroin users usually started with other substances before beginning that kind of regimen, glue sniffing was in no way an endgame drug. But this analysis was only slightly more effective. Later in the decade, Dr. Joel Fort would compare the "stepping-stone mythology" of gateway drugs to the federal government's problematic domino theory of foreign policy, which loosely justified its actions in Vietnam based on fear rather than logic. Proof of cause and effect was largely absent in descriptions of stepping-stone drugs. But while glue as a gateway to marijuana, cocaine, and ultimately heroin was certainly fretted over by parents and doctors alike, the more immediate fear of spread (the fear that would ultimately affect any contact trace of the problem) was the fear of spread to other products readily available in the American household. Gasoline had been a problem early in the century, as had turpentine and household cleaners. In 1891, the American Medical Association publicly worried about children sniffing ether. "Little schoolboys were seen clubbing their pennies to purchase a vial of the exhilarating fluid."[30] In at least two cases, the result was death. Ether was no longer in the majority of American homes, but sniffing model airplane glue only reminded children that there were still plenty of things to sniff around the house to give them the broader effects of ether. A 1964 study of Bexar County, Texas, demonstrated that lighter fluid sniffing had become a significant problem, for all of the same reasons that spurred the use of glue. Aliphatic hydrocarbons, and naphtha in particular, produced intoxicating effects described in markedly similar terms to those of sniffing glue—stupor, dizziness, "pleasurable oblivion."[31] But lighter fluid also caused headaches, nausea, irritability, and damage to the mucous membranes. It created delinquent behavior, something akin to addiction (in much the same way glue was producing something akin to that loose definition of addiction), and poor school and intelligence quotient performance. There were almost always problems at home from the boys who used. Latinos and whites were the principal sniffers.[32] The formula was exactly the same, but the product was different, and the contagion was something else entirely. The more answers researchers came up with, the more the contact trace map changed.

Assuming that gasoline, lighter fluid, paint thinner, nail polish remover, turpentine, or paint were ever absent from the paradigm was problematic. They were still in the house. They were producing similar effects. And the problems associated with the use of such products were comparable to those of sniffing glue. All, for example, were fat-soluble organic solvents, passing the blood-brain barrier with undeniable speed. They all seemed to alter myelin, the substance covering the axon of neurons, thus hindering the function of the nervous system. Researchers even attempted to simulate "inhalation psychosis" in a controlled environment, using paint thinner as the reactive agent. It took between fifteen and twenty-five "drags" to start to see visible effects of a high. About ten more drags achieved full-blown psychosis, which tended to dissipate in about an hour. Electroencephalographic tracings on the test subject demonstrated a marked slowing of the brain waves, corresponding to textbook delirium found in abuse of alcohol and barbiturates. But then, in other studies, delusions and hallucinations were obvious, creating a legitimate psychosis on par with schizophrenia. It was assumed, as it was assumed in the 1930s, that such was the creature of "the basic ego-weakening process" that came along with such use. Delusions and hallucinations were "more likely to occur in those whose psychic economy was in precarious balance prior to the delirium."[33] And since not all adolescents had a working knowledge of their "psychic economy" before use, the particular state that drove each user was going to be different, based largely on what they were like before use.[34]

And when life had prepped them properly, argued Frederick Glaser in early 1966, they were going to crave it, whether the term "addiction" was ever settled or not. In both Ohio and Texas, for example, instances were reported where parents literally chained up children to prevent inhalation, neither able to sufficiently bind the kids to keep them from escaping back to their favorite fat-soluble organic solvent. So again the term "addiction" was beside the point. So too, for Glaser, was long-term organ damage. "Whether or not it is found that a certain amount of direct toxicity to various organ systems does exist, the practice of inhalation is dangerous and occasionally fatal."[35]

Glaser's portrait of sniffers was similar to others reported throughout the early 1960s, but combining this later study of glue sniffers with other kinds of drug abusers was instructive. Sniffers were adolescent males, for the most part, and there was a high incidence of Latinos who used. Black kids sniffed glue in various places, but their numbers seemed generally to be lower for sniffing glue and other solvents. Still, "the percentage of Negro admissions for narcotics addiction to the USPHS Hospital at Lexington [Kentucky] has fluctuated from 8.9% in 1936 to 52% in 1955 to approximately 41% in 1964."[36] Preferences for certain intoxicants might have been generically different, but the preference for intoxicants was relatively universal, and if that was the case, minority status couldn't be a fair predictor of possible use. In addition, while inhalant users almost never had corresponding problems with heroin and cocaine (at the time of use—glue's status as a gateway to later use of such drugs hadn't been qualitatively established), they did, according to several studies, take part in tobacco and alcohol abuse. Of course, that wasn't an indicator of anything for Glaser. The problem was more complicated than that. There was no valid way to presuppose use of one substance based on use of another. To that end, Glaser told the story of an eleven-year-old Australian boy. "He said he had been eating rubber erasers at school for the last four years, had run out of them, had 'wanted something nice to do,' and so decided to sniff petrol, of which he had always liked the smell."[37] The final hindrance to any developmental contact trace was the arbitrary nature of the thinking of children. Often, there didn't have to be a reason.

Still, broken homes, antisocial behavior, and below-average intelligence all seemed to combine with sex, age, and race to at least provide indicators of possible use. As did the sense of smell. Glaser, like many others before him, retrod the boards of psychoanalytic debates about the sense of smell. "Transmutations of coprophilia, an anal stage phenomenon," was still considered a viable diagnosis. There could be a general feeling of eroticism, masturbation, and fantasy. Users could "create a wish-fulfilling situation which allows a certain amount of instinctual discharge—a discharge which would not be permitted in the existing circumstances of external reality—and which also corrects and modifies that reality in the imagination."[38] Such psychoanalytic paradigms only added to the worry over potential use and its possibilities to corrupt but also provided a symptomatology that couldn't be traced by any categorical epidemiological portrait. By the 1960s, illusions that only demented children masturbated had faded away. And what young experimenting masturbator wouldn't want to enhance the fantasy of the proposition or alleviate a measure of shame attached to the act? Such psychosexual impulses were nothing if not the white background upon which all of the contact trace dots appeared.

Seeing inhalants as a broader group, however, could provide some useful results. Gasoline sniffers, for example, were likely to be loners, while "glue sniffing is usually practiced in groups or 'gangs.'"[39] Other studies had portrayed glue sniffers as lonely isolated children, but a study by Columbia University researchers Sylvan Bartlett and Fernando Tapia argued that a tendency toward isolation really didn't matter. The tendency of glue sniffers to practice in groups was pragmatic. Several tubes of model airplane glue were squeezed into a paper bag to create the high concentration of vapors necessary for intoxication. But the vapors diminished as the glue hardened. "Thus, several persons contribute glue which is squeezed into a bag and the bag is passed from one to the other before the solvents have evaporated."[40] So while the predisposition of most adolescent glue sniffers might be one of lone activity, the process of use, according to Bartlett and Tapia, necessitated the presence of a group. In addition, the majority of gasoline inhalation cases came from rural areas, whereas glue-sniffing incidents came predominately from urban areas.[41]

Of course, the pragmatic need for practicing with several sniffers wasn't really a pragmatic need at all. Tubes of model glue were cheap, and if children lived in an area where laws against minor sale hadn't been passed, there wasn't any need for fellow sniffers. Group use did seem common, but it didn't seem to be any real necessity. And while the furor over glue sniffing clearly came from urban areas, children made models in all parts of the country. The Bartlett-Tapia assumptions were useful aids in furthering the glue-sniffing portrait. Urban adolescent boys from broken or troubled homes who tended to be white or Latino, or who demonstrated delinquent behavior, failure in school, and low intelligence quotients, could have tendencies toward withdrawal from society while still ultimately finding solace in the more limited society of friends and fellow glue sniffers. The patterns of abuse and the need for ever heavier amounts of glue to produce the hallucinogenic effects that sustained sniffers were also valuable insights.

But the academy's research, its overarching effort at quantifying and qualifying a problem that seemed to be growing by the day, seemed to show the same limitations as the lurid headlines. Progress was slow. Researchers crept incrementally toward the kind of data that could lead to an epidemiological quarantine, but as they plodded slowly toward the finish line, the finish line just moved farther and farther away. In some cases, as with their problematic reworkings of the nature and definition of addiction, the researchers were pushing it away themselves. Regardless, the spread of adolescent glue sniffing seemed to be moving much faster than the research. Kids didn't require peer review before huffing and didn't read the dire warnings coming from the nation's hospitals and universities. As scientists honed their definitions of the contact trace dots, the number of dots continued to multiply beyond their reach.

# The Devil in the Room

"THE DEVIL IS COMING in the window to take me away!" the frantic boy screamed to anyone beyond the thick door who might be listening.[1] He was thirteen and had earlier been detained in the juvenile hall for trying to rape a young girl at an Oxnard, California, movie theater while under the influence of model airplane glue. In Imperial Beach, just outside of San Diego, a sixteen-year-old sniffer beat his parents and younger brother. A group of glue-sniffing youths in San Dimas fired fifteen shots into a building. A frightened California began debating a bill to stanch such behavior in March 1963, but controversy over the inherent problems of small business regulation and fear of frivolous legislation kept the bill from passing. Such violent cases, however, became harder to ignore. The devil was coming to take California children away. Finally, Senator Thomas Carrell of San Fernando introduced a new bill in March 1965. It classified model glue as a poison and prohibited its sale to minors. With the problem metastasizing all around them, the legislators acted swiftly the second time around. The bill became law in July 1965.[2]

California itself was a study in paradox—the state that housed so much of the counterculture was also home to reactionary drug laws. By 1965 Ken Kesey and his band of Merry Pranksters had already traveled from La Honda to the 1964 World's Fair. Students at the University of California at Berkeley had already chained themselves to administration desks and started a student movement for free speech. By November 1965, when Kesey would hold the first of his famed "acid tests," the state was already prohibiting model airplane glue sales to minors, but it was also regulating heavier narcotics, emphasizing punishment rather than treatment. Even a first offense for marijuana possession garnered the convicted person up to ten years in prison, with no parole for the first year. For every Haight-Ashbury, there was a Richard Nixon or a Ronald Reagan. It was a state that bred some of the most significant renunciations of received culture but also some of the most significant renunciations of the renunciations.[3]

The hole in the California model airplane glue legislation, however, was that it didn't account for users who might not be children. California was crafting a juvenile delinquency law, not a drug bill, and this mattered (see chapter 8). As the bill was becoming law, for example, a forty-one-year-old baker died of an intra-alveolar hemorrhage of the lungs caused by toluene irritation. A twenty-four-year-old lapsed into a coma after sniffing glue and died before reaching the hospital. The autopsy report was familiar. The death was "compatible with toluene intoxication from glue sniffing. Intra-alveolar hemorrhage, lungs, minimal. Thickening basement membrane, bronchial mucosa."[4] In Denver, a twenty-one-year-old man, Arthur Garcia, was arrested after robbing a local drugstore of its model airplane glue supply. He claimed to have been sniffing glue for eleven years. Three other twenty-somethings were arrested by Denver's Vice Bureau after being discovered huffing glue in a downtown theater.[5] There were plenty of adults sniffing glue, as well, and there was no restriction on their ability to purchase it, even though death was clearly a possible consequence of such use.

The deaths were mounting. Oregon, Wisconsin, Kansas, Massachusetts, California, Pennsylvania, and Virginia all reported deaths related to toluene poisoning. But the deaths that happened ancillary to glue sniffing were far more disturbing. A Los Angeles boy was beaten to death at a glue-sniffing party. A South Carolina boy, his friend reported, "kinda went crazy and acted like he was a bird. I left, and when I got back my friend had shot himself in the head and was dead."[6] It was clear that such behavior, whether it caused death or not, was completely unpredictable. "Imagine the consequences," said one police official in 1962, "if the subject experiences hallucinations while driving, working with tools, or if he is placed in any situation where he could endanger the lives of others or harm himself."[7] By 1964, no one had to imagine. A twelve-year-old boy in San Francisco robbed a florist. Another beat a paperboy and robbed him.[8] Robberies, violence, and accidental falls from rooftops were becoming more and more commonplace.

So glue sniffing was clearly a social problem. "The youth who starts out with glue-sniffing and ruins a career with narcotics," cried the *Christian Science Monitor*, "is both victim and advocate of the criminal element." Oberia D.

Dempsey, a minister at Harlem's Upper Fifth Avenue Baptist Church and an antidrug activist in the community, estimated that sixty thousand of Harlem's four hundred thousand residents were heroin users, and that another forty thousand were being groomed for the practice by smoking marijuana and sniffing glue.[9] One heroin addict for every 150 residents seemed like an incredibly high estimate, but the gist of Dempsey's claim was that substances like glue took hold of their users and created in them if not an addiction at least a predilection to finding a high at whatever the potential cost.

And why not? Such was the maker of social problems. For all of the academic debate about the nature of addiction, it was clear that there was something similar to dependency in many users of model airplane glue. A sixteen-year-old Fresno, California, boy began crying and screaming frequently and without purpose. His grades diminished. He attempted suicide. What doctors assumed to be a brain tumor turned out to be a chronic dependence on model airplane glue, lasting between six and eight months. After a five-week forced hospital stay without glue, everything returned to normal. His grades improved. The crying stopped. There was no more screaming, no more suicide attempts.[10] As with all social problems, model glue had the power to hurt families and individuals, even when specific violent crimes weren't being committed.

Meanwhile, the hobby industry wasn't cowering, but it was certainly feeling the weight of such turmoil, a weight that bore down on everyone from shop owners to manufacturers. In Parsippany, New Jersey, for example, hobby store owner Richard Palmer began hiding his glue supply under the shop's counter and voluntarily limited his sales to two tubes per customer, fearful that the next fatal glue-sniffing binge would start with a product that came from his store. "We spend years trying to build up interest in a creative hobby," Palmer complained, "and then some idiot who hangs out on a corner gets an idea in his head and this wrecks our plans." But in a broader sense, the hobby industry's $450 million in annual sales were threatened at every link in the chain. "We're going to have to face up to it," said William L. MacMillan, the HIAA's executive director in late 1962, "and either change our products so they can't be sniffed or at least educate communities on what to do if they're faced with such a problem." The association claimed to have spent $250,000 that year to combat the growing menace.[11]

But the menace wasn't easily combated. Much of the HIAA's money went to Applied Biological Sciences Laboratory in Glendale, California, which was trying with little effect to decrease the intoxicating effect of model glue. Individual companies also made similar efforts, and with similar results. Ambroid Company, for example, based in Weymouth, Massachusetts, tried to find bonding agents other than toluene, but none of them could produce the quick-drying effect necessary for model-making. Testor claimed to have changed formulas several times throughout the early 1960s, but with little result. DuPont thought the only way to stop the trend was to ward off would-be sniffers with odious smells but feared that doing so would stop law-abiding model-makers from continuing the hobby. Still, for all the effort the companies put into solving the intoxication problem, they vigorously defended the industry. Testor president Charles Miller reminded people that "you can use almost any product you get from petroleum and produce the same intoxicating results."[12] If media coverage was generating the epidemic (which in large measure it clearly was, considering that the development of new "outbreaks" followed hyperbolic coverage of the growing problem—such is where the "idiots" got their ideas), then Miller's statement wasn't exactly socially responsible but it did have the benefit of being true. The American Medical Association estimated that there were as many as 250,000 potentially harmful household products on the market in the early 1960s.[13] Miller was comfortable in the knowledge that this wasn't the first time teenagers had made corporations the accessories to borderline criminal behavior.

In the late 1940s, for example, teenagers began commandeering Benzedrine nasal congestion inhalers, tearing them apart, and chewing on the medicine-soaked cotton inside. Companies worked to make stronger inhalers that couldn't be opened but finally produced a new compound in 1949 that would effectively clear nasal passages without getting kids high.[14] Focusing on glue's smell and deflecting blame to other products were the red herrings comparable to making stronger inhalers. Only finding a safer chemical was going to do the trick.

A safer chemical had yet to appear by the summer of 1964, however, and New York City, awash in narcotics addiction, faced what police commissioner Michael J. Murphy called "a new scourge," glue sniffing, which he described as being closely related to narcotics.[15] Murphy surely knew that the scourge wasn't new in the summer of 1964, but his description of the problem was emblematic of people's troubled thinking about the products getting their children high. Glue was closely related to narcotics, it was part of the narcotics problem, feeding a desire for euphoric effects that ultimately led to other drugs. Still, being closely related to narcotics means that model airplane glue wasn't a narcotic, wasn't quite the same thing as cocaine or heroin. Glue's inability to meet standard categorizations created a nebulousness that allowed children to keep using it while adults tried to come to grips with just what it was exactly. Meanwhile, calling it a "scourge" only exacerbated public fears.

Whatever model airplane glue was, officials were becoming more sure that something needed to be done to regulate the problem. In response to the death of a thirteen-year-old Brooklyn boy, who was sniffing glue by the Gowanus Canal, fell in, and drowned, Alicia R. O'Connor, deputy commissioner of the Nassau County Board of Elections, urged the passage of a state law to prohibit sales to minors. O'Connor was running for family court judge and knew that anyone who was staking a claim to a position involving the health and morals of minors would need to stand publicly against the new menace, whether it was technically a narcotic or not. And she wasn't alone. Late in 1964 the Philadelphia city council proposed its own law to ban glue sales to minors.[16]

Even as they did, sniffers were finding other substances. Prisoners at Rikers Island were caught sniffing mace. Twenty delinquents from the Montana State Vocational School for Girls rioted and broke windows after sniffing fingernail polish remover. Throughout the first half of 1965, two New York teenagers died and four others were hospitalized after sniffing Carbona cleaning fluid.[17]

Meanwhile, the question of addiction as it related to glue and other toxic solvents was entering the mainstream. Glue sniffing, marijuana, pep pills, all could create dependencies, argued producer Jack Landau, who created a television documentary on the subject. "Addiction and pre-addiction are becoming quite alarming, particularly since the problem is no longer confined to the slums or the low-income bracket. 'Nice' kids are now involved." Officials in Nassau County, New York, formed a Narcotics Task Force to fight addiction, including glue sniffing. "It's a damn shame," said Palmer Farrington, presiding supervisor for the town of Hempstead. "I can't help but feel that something should be done about this problem." That day two high school students from Oyster Bay were arrested for sniffing glue with bags over their heads while in a car, a car that officials assumed the teenagers had been driving.[18]

This was in late February 1965. In early March, Washington, DC, officials claimed to have the problem under control, arguing that the regulation against sniffing glue had its desired effect. "We've had a few cases since [the regulations]," said deputy police chief John Winters, "but right now it's no longer a problem." But *Washington Post* columnist George Lardner Jr. didn't think that laws were the reason for the diminishing statistics. The Washington statute banned sales to minors and required merchants to keep records of sales. "Certainly it's doubtful that many merchants have been paying the least bit of attention to what the Commissioners ordered." Winters said that no reports had arrived claiming that hobby shops weren't properly keeping records, but the police had never gone to stores demanding to see them. Lardner assumed that kids had "simply gotten tired of the fad," but he was frustrated that the regulations weren't being enforced. District officials "seem to be turning their noses the other way. They could look pretty silly if glue-sniffing comes back into vogue."[19]

It still seemed in vogue everywhere else, and by June, the case of the Oyster Bay teenagers had reached the Nassau County District Court. The judge, Beatrice S. Burnstein, however, threw out the boys' case. Albany never responded to Alicia O'Connor's call for statewide legislation, and without it Burnstein ruled that the case had no merit. The boys had been charged under a statute that prohibited injury to personal property, disturbing the public peace, disturbing public health, or outraging public decency. "Courts may fill the interstices of vague and ambiguous statutes, but they cannot enact a statute, particularly in the field of crimes, where the Legislature has failed to act." Glue sniffing might have had several commonalities with the broader statute under which the boys were charged, but the statute never made sniffing glue a crime. Certainly glue sniffing did "offend against moral standards and reflect, perhaps, a drive by a handful of young people, bent either upon escape or rebellion, to experiment with new forms of antisocial adventure," but this didn't mean it was against the law. In the same line of thinking, glue sniffing was "self-evidently injurious to health"—but only to the health of the user.[20]

This is not to say that Burnstein was somehow in favor of allowing children to huff at will. "Glue sniffing has become a national problem, and legislators in all States have been alerted to it." But it was not listed as a narcotic by the state of New York. Burnstein reviewed pending New York legislation designed to attack the problem and worried publicly that it only restricted sales. "There are no penalties provided for the possession of glue," and without such penalties in place, no glue-sniffing arrests would hold up in court. "To be sure," she argued, "penal sanctions alone will not reach the problem, but it must be evident now that the absence of sanctions either encourages, or creates a climate for, the commission of antisocial acts."[21] It was the most public statement to date that glue sniffing needed regulation. It needed deterrence. It needed a law in New York.

But "penal sanctions alone will not reach the problem." One letter writer to the *New York Times* suggested lowering the beer drinking age from twenty-one to eighteen. That would, he argued, keep younger drinkers from hard liquor, which was easier to steal than beer, and it would also entice would-be glue sniffers away from the practice toward something more socially acceptable. Such a suggestion was more than unlikely, but there were still plenty of nonlegal avenues for helping glue sniffers. Poison control centers had been growing as part of state health departments since the 1950s. As of 1965, New York's poison control center had fielded more than 117,000 calls, most dealing with complications from household products. The majority of the cases were the results of accidents, but abuse of substances like model airplane glue was a recurring problem with which Poison Control dealt regularly. But it wasn't easy. "There is some risk in drawing conclusions," said Harry Rabin, Poison Control's technical director. "Sometimes doctors and hospital personnel prefer others to make decisions for them."[22]

Prompted by the Denver juvenile court, which had taken the lead on glue-sniffing research since the epidemic's inception, the US Welfare Department's Office of Juvenile Delinquency and Youth Development provided a $57,451 grant to Denver for an experimental treatment program for glue-sniffing addiction. Boys in the program would undergo "therapy in four groups, under four different methods."[23] Then doctors would evaluate which among them was most effective in combating the problem. "The boys are being taught to adjust to their home surroundings, which often are poverty-stricken, and to work to improve them rather than resorting to glue-sniffing as an escape."[24] The grant was significant for two reasons. One, the action by the Welfare Department was the first federal effort to combat the glue-sniffing epidemic, signaling that the problem had grown so broad in its scope that the federal government felt the need to respond. By the summer of 1965, of course, the Great Society was underway, and Lyndon Johnson had proved far more willing to enact laws spurring government intervention into the domestic economy in his "unconditional war on poverty," but that didn't make a half-million-dollar expenditure any less momentous for a federal government that had never acted on behalf of the growing concern over model glue.[25]

The second reason the grant was so significant was its donation of funds for the creation of a treatment program for addiction. The word had been bandied about for years in association with sniffing glue, debated as to whether or not it applied to the use of such substances. Certainly the cavalier use of "addiction" as a concept related to model airplane glue (see chapter 4) didn't ensure that the grant was a definitive statement from the federal government on whether such terms were accurate or misleading. But now the federal government was spending its money not on corporate regulation of hobby corporations like Testor, and not on prevention programs for youth at risk. It was spending its money on addiction.

It seemed like an odd choice, considering the immediacy of most of the lurid headlines involving glue sniffing. A kid jumps off a roof or tries to beat up a marine or falls into a canal and drowns. Stories like these didn't result from long-term overuse. They were the result of immediate complications with one specific use. Denver, however, had taken the lead on the problem, and if the research tended toward the assumption that model airplane glue was at the very least habit-forming, then prevention programs were going to trend toward that way of thinking. Everyone else, meanwhile, was left with the frightening immediacy of glue's short-term consequences.

Despite all the science that said black adolescents were less susceptible to the epidemic than most, Harlem was knee-deep in the frightening immediacy of those consequences. As David Courtwright has noted, black families have historically been more prone to family disruptions—in the form of divorces, separations, and abandonments—than have white families. While such statistics have understandably been skewed by poverty, racism, and the consequences of those factors in American life in the post-Reconstruction United States, those disruptions did have the consequence of creating a higher percentage of vulnerable black youths.[26] In early 1965, Harlem Youth Opportunities Unlimited (HARYOU) and the Associated Community Teams of Harlem (ACT) began lobbying police for demonstrations designed to show the danger of sniffing glue and other household products and medicines, and how those behaviors could serve as gateways to addiction to more serious drugs like heroin. Kenneth Clark founded HARYOU in 1962 as an education and employment agency in Harlem. HARYOU lobbied Washington for education spending, it produced a report on the Harlem riots of 1964 and created educational programs designed to keep such things from happening again. That year, HARYOU merged with Adam Clayton Powell's ACT. The combined HARYOU-ACT clearly understood that narcotics addiction was the principal instigator of failing homes and increased criminality, which in turn eroded the educational opportunities and abilities of Harlem students. By petitioning the New York police department to help teach children about the dangers of the seemingly innocent forays into sniffing glue or drinking cough medicine, the group was hoping to kill two birds with one stone.[27]

The police responded. In the summer of 1965, they began a series of demonstrations on select street corners throughout Harlem, a mobile unit in tow complete with an exhibit about the known dangers of model airplane glue and other toxic chemicals, as well as the harder drugs just waiting down the road. As of the fall, police programs planned for locations near schools to ensure the heaviest traffic for their juvenile audiences. Washington, DC, took a more formal approach, as the American Society of Humanistic Education held a roundtable of leading childhood experts on the topic of "New

Answers to the Problems of Children." Glue sniffing and its antecedents—delinquency, dropouts, underachievement, and sexual experimentation—were the subject of the workshop's intense focus.[28]

But authorities in other locales were far less interested in examining and fixing the problem, choosing instead to pretend it didn't exist. In the summer of 1965, as police began teaching demonstrations on the streets of Harlem, William McMyne, mayor of Wharton, New Jersey, suspended his police chief, Abe Hocking, for telling newspaper reporters about "incidents in which teen-aged girls had turned their baby-sitting jobs into rowdy parties that involved beer-drinking, glue-sniffing, one knifing and larceny." The community was incensed at the suspension, the Civil Defense Board threatened to resign, and the move clearly backfired on McMyne. The mayor more than likely feared that the release of such information would sully the town's reputation and ultimately hurt local business interests, but it seemed to be a relative exception to the declamation of falling skies coming from public officials in other parts of the country. The public clearly felt it had a right to know about the glue-sniffing epidemic reaching Wharton.[29] Hocking was reinstated, but two elements of the story were lost in the furor over McMyne's attempted cover-up. First, it was clear that much of the country's glue-sniffing problem was going unreported, either to police or to the press, leaving officials with only hypotheses as to the actual casualty numbers. Second, the problem in Wharton was female. Against all evidence from any legitimate contact trace, it was babysitting girls who had masterminded the glue-sniffing parties. It was girls involved in thefts and stabbings. And so the specter of glue-sniffing grew larger—media cover-ups and girl masterminds only broadened what the problem could be.

By autumn the problem had broadened to the upper-class suburb of Darien, Connecticut. Westport, Stamford, and Bridgeport all had similar problems. In 1964, only 28 percent of narcotics cases in southern Connecticut involved low-income families, and that number had continued to decline through the first three seasons of 1964. Teenagers from the more affluent homes of the already affluent region had access to cars, which gave them access to New York City, which gave them access to the products and influences that ultimately hurt places like Darien. Police there had made arrests for glue sniffing, barbiturate use, and marijuana, all of the accused being students at Darien High School. "In the suburbs the over privileged kids are getting disassociated from the tools of living," said one social worker in the area. "They're not being taught to fend for themselves, to make their own way."[30]

Just as in Wharton, Darien officials stumbled quixotically throughout late November 1965 to keep the story from spreading. "I have nothing to say whatsoever," said Darien High's principal, Stuart Atkinson. "The newspapers are trying to crucify these children." The notion, however, that the Darien stories didn't have merit seemed far-fetched. A better protest came from a local bank executive, who asked, "Why pick on Darien? The same thing happens in New Canaan or Greenwich and no one cares." He was right about the fact that Darien wasn't the only affluent town affected by the problem, but people did care. In early December, seven teenaged boys and two teenaged girls were arrested in Bridgeport in connection with a glue-sniffing party in late November. Two of the arrested were escapees from the Meriden School for Boys, but the others hadn't been considered problem kids.[31] Such was the nature of the glue-sniffing menace. Emphasis on low-income families and neighborhoods neglected the fact that wealthier children had far more time, means, money, and model kits for abuse and intoxication.

Both ends of the income spectrum were affected, but so too were the middle-class kids in between them. Long Island had seen its share of glue-sniffing trouble throughout the first half of the decade, and on the same day that Bridgeport made its arrests, Long Island detectives captured more than twenty dealers as the result of a year-long investigation into narcotics trafficking among teenagers. Model airplane glue was listed along with peyote, marijuana, and heroin as the principal drugs sold. Theodore Wallace, a fifty-one-year-old owner of a stationery store, sold model airplane glue "by the caseload" to one teenager, who would then distribute it to friends and buyers. Glue was found alongside syringes and sawed-off shotguns. "We have to face up to what has happened here," said Northport mayor Gilbert H. Scudder.[32] "This is unlike any problem in the past. Shock is not enough to solve it."[33]

It was unlike any problem in the past. Glue trafficking alongside peyote and heroin seemed unquestionably to place glue in the category of narcotics, and the raid also demonstrated that glue was a severe middle-class problem. It also allowed officials to tie the culture of fear and demagoguery associated with hard drugs to model airplane glue.[34] When sawed-off shotguns became involved, the headlines got bigger, the jeremiads more severe, and the stakes higher.

Glue, however, needed no weapons to present its range of problems. In early 1966 two New Jersey adolescents suffered 15 percent brain damage and what officials described as "total sterilization" after sniffing model airplane glue. But even without such effects, glue was clearly damaging American junior high and high school campuses. As hallucinogenic drugs became more popular in the mid-1960s, pushed by advocates like Timothy Leary and sensational accounts like Tom Wolfe's *The Electric Kool-Aid Acid Test*, glue took its place as the youthful, available version of drugs like mescaline and LSD. By the mid-1960s, use of hallucinogenics and dependence on substances like model airplane glue and marijuana had ballooned throughout the country, and high school dropout rates had grown, as well. The connection between the two phenomena wasn't ironclad. Most students claimed to drop out because of rejection by fellow students, dissatisfaction with courses, failing grades, or the need to work. But in many cases drugs did play a role.[35]

In March 1966, John Hersey's novel *Too Far to Walk* appeared, telling the story of existentially troubled sophomore collegian John Fist, who falls in with a bad influence named Chum Breed, a character who had begun sniffing glue at age fourteen. By college, Breed had advanced to LSD, pulling Fist into his habit.[36] That was what could happen. Glue use in adolescence could lead to hallucinogenic drug use in college, thereby derailing promising futures and beautiful minds.

In response, officials began seeking new money for new programs. It was a common practice that critics saw as fundamentally disingenuous. "Self-serving statements by drug policemen and their agencies or other government departments, and overblown accounts of isolated individual

Dropouts, in a national survey, cited adverse experiences—falling grades, rejection by fellow students and dissatisfaction with courses—as the most common reason for leaving school.

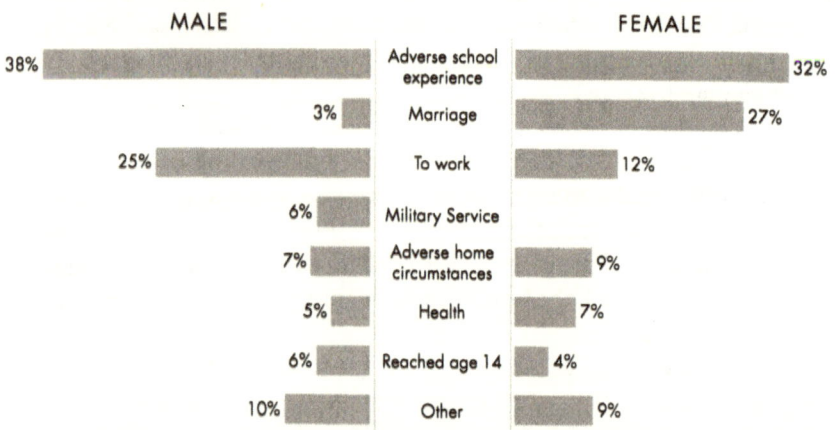

THE REASONS STUDENTS DROP OUT: This *New York Times* diagram differentiated the most common reasons that male and female students dropped out of school. Source: Fred M. Hechinger, "Education: Drugs—Threat on Campus," April 10, 1966, 151.

or group drug experiences," argued Dr. Joel Fort, "have contributed greatly to the incorrect impressions existing in our society about drug use and drug problems." In Richmond, Virginia, corrections officials lobbied Congress for a new youth correction and rehabilitation facility after witnessing several teenage boys return repeatedly to juvenile halls across the state for sniffing glue and other offenses. Meanwhile, Harlem was still begging for help from city officials in 1966. "We have had studies and more studies on what should be done in Harlem," said Donelan J. Phillips, president of the Consolidated Tenants League, a black tenants' advocacy group. "There are many, many programs 'on paper,' but nothing is being done." Sniffing glue, harder drug use, poverty, homelessness, and dropouts all plagued Harlem, but while HARYOU-ACT and other groups were still receiving federal money, the city itself was slow to respond to problems in black neighborhoods.[37] As Jack Landau had hinted the previous year, official concern only showed itself when "nice kids" were involved, and the nice kids were white.

And so locals began trying to fix the problem on their own. In Newark, New Jersey, Reverend William Iverson created Bill's Rough Riders, a restaurant and hangout joint directly across from West Side High School. He used the restaurant as a staging ground for a ministry devoted to stopping teenagers from sniffing glue, snorting cocaine, and shooting heroin. "Some of the kids can't talk with their parents," he explained. "They feel unloved. We give them a chance to talk and let stuff out. When they see you'll listen, they just love to tell you what they think, to confide."[38]

Everyone knew that glue was affecting any and all types of adolescents, not just the "nice kids" and not just the students at West Side High. And if that type of pharmacology was color-blind, so too was the psychopharmacology that sought to treat it out of existence. Doctors were "very much concerned about recent developments in multiple drug dependency and use among young people in the suburbs of large cities here and in Europe," Dr. Henry Brill told an audience at the Fifth Congress of the Collegium Internationale Neuro-Psychopharmacologicum. "It is a worldwide problem." And the reason it had gone global was the ease with which people could communicate in the twentieth century. "Glue-sniffing is reported here, and it spreads to England in a short time."[39] Communication and knowledge were the lines connecting every contact trace dot, and with the kind of scare-headline reporting that had taken place over the first half of the decade, communication and knowledge were everywhere.

Where communication and knowledge seemed to break down most readily was within American homes themselves. There was, for many writers of the day, a vast disconnect between parents and children—and an even more dangerous denial by parents that any such disconnect existed. "I am angry at parents who are afraid of their children, who indulge and fawn upon them to gain their goodwill," went a typical jeremiad, this one by author Dorothy Gordon. "I am angry at our sex-geared society that preys upon our youth—at the movies, at the ads, the mass media that wean children away from childhood pleasures, throwing them into a synthetic adulthood before they have even realized a meaningful youth."[40] Gordon's declamation was typical of a spate of books published in the mid-1960s crying over the communication gap between parents and children and prophesying American doom because of it. "Matters could not be much worse than they are now," wrote another such author, Thelma C. Purtell. "The night has become very dark indeed. If this presages a dawn, it will be only because all of us, as parents, have worked and willed to bring it."[41]

Prophecies of doom weren't new to the 1960s, but the spate of books on teenage morality and its relationship to parenting signaled a significant change in some representations of the glue-sniffing epidemic. Books like those of Gordon and Purtell didn't classify model airplane glue with other toxic household products. Neither did the Collegium Internationale. Neither did the reformers in Harlem and Newark. Instead, model glue was classed with other elements that were destroying the moral fiber and potential success of American youth. Glue was placed in comma-delineated sequences with cocaine, heroin, sexual experimentation, truancy, and vandalism. Considering the development of the epidemic and its place in the adolescent and teenage spheres of influence, this seemed like its proper place. It wasn't as statistically dangerous or addictive as heroin. It didn't come with the incumbent possibilities of disease or pregnancy as did sex. But glue's tendency to create isolated, intellectually and emotionally stunted kids with a predilection toward delinquency clearly placed it among other problematic teenage habits—and particularly among those that caused rifts between parents and children.

As 1966 began to wind to a close, it was clear that better descriptions were helping. HARYOU-ACT received more than $11 million in aid in its first two years of operation, much of that in federal money. In October 1966, US Welfare Department commissioner Ellen Winston announced that the federal government was adding an additional $44,923 to the Denver juvenile court's experimental glue addiction treatment program. And even the Washington, DC, glue-sniffing statute, declared unnecessary by city police and ridiculed by columnist George Lardner, was still working for the District. Not only were police still catching and prosecuting glue-sniffing offenders, but the juvenile court's youth aid division was using the law to process adolescent marijuana offenders, as well.[42]

That same month, the federal government made LSD illegal, despite the best efforts of advocates like Timothy Leary, who the month prior had formed the League for Spiritual Discovery to make a First Amendment religious claim against the ban. Still, while the federal government was regulating hallucinogens and supplying grant money to model airplane glue studies, it was doing so separately. The Welfare Department's grant and the legislature's law were different entities doing different things. Because one was a child's problem, one a teen's, the two responses were never seen as mutual.[43]

On November 3, 1966, the federal government went further in aiding the child's problem, modifying the Federal Hazardous Substances Act "to ban hazardous toys and articles intended for children, and other articles so hazardous as to be dangerous in the household regardless of labeling." The Child Protection Act of 1966 was a watershed, but at the same time, as with so many other elements of the official fight against the epidemic, its blind spots circumscribed its better intentions. The law banned any "toy, or other article intended for use by children, which is a hazardous substance, or which bears or contains a hazardous substance in such a manner as to be susceptible of access by a child to whom such toy or other article is entrusted." Importantly, however, the law exempted "articles, such as chemical sets, which by reason of their functional purpose require the inclusion of the hazardous substance involved, and which bear labeling giving adequate directions and warnings for safe use and are intended for use by children who have attained sufficient maturity." Model airplane glue was never mentioned, but clearly it fell within the exempted articles. The hazardous substances in model glue were only potentially hazardous. A chemistry set, for example, could cause significant damage through accidental mistakes, through good intentions. Glue's damage only manifested itself when the user's intentions were more sinister, when such use was intentional. That being the case, the government's interpretation of wrongdoing in cases of destruction through glue sniffing fell solely on the user. Glue was only hazardous to people who wanted it to be hazardous. And so, even though the country was knee-deep in the fight against glue sniffing, model airplane glue was excluded from the act.[44]

It was a well-intentioned, beneficial attempt at protecting children from their toys, but its nearsightedness on glue sniffing left it impotent to protect them from one of the most immediate dangers they faced. Even with such beneficial activity, however, no amount of legislation, no amount of money had lessened the number of scare headlines, the number of juvenile offenders, the number of frantic boys frightened by imaginary devils. More needed to be done. And so, in early 1967, the leading experts in the field met at the epidemic's ground zero, Denver, to figure out what needed to come next.

# The Glue Summit

DENVER WAS COLD IN January, the kind of cold that cut through whatever well-intentioned garments the researchers used as they stepped off the plane at Denver International Airport in 1967. Since the initial spread of the epidemic, representatives from hobby industries, health agencies, and law enforcement had met each year to discuss the inhalation of model airplane glue and other household solvents and potentially addictive compounds. There were acts of worried hand-wringing and declamations of peril. There were recitations of the literature that had appeared over the course of the past year. There were discussions about possibilities for limiting the problem, through either legal remedies, parental responsibility, or industrial replacements for solvents that could prove either intoxicating, addictive, or both. At the fourth such conference, in Berkeley, California, in January 1963, researchers touted the possibility that Freon could be a potential replacement, being that it was "inert physiologically," didn't produce intoxication, and was nonflammable.[1]

Freon, it turned out, wasn't really a viable substitute for toluene, but there were problems with law enforcement as a way out of the problem, as well. At the epicenter of the epidemic (Denver), for example, in 1960 officers arrested 30 sniffers. In 1961 they arrested 134 sniffers. In 1964 they arrested 164. Arrests had increased fivefold in four years.[2]

Such conferences continued, with overdeveloped intentions and underdeveloped results. In 1966, however, the Denver juvenile court and the US Department of Health, Education, and Welfare's Office of Juvenile

Delinquency and Youth Development planned a more expansive, substantial conference in January 1967. From January 23 to 25, criminologists, behavioral scientists, sociologists, and all sorts of delinquency workers—from volunteers to youth workers to police officers to therapists to judges—descended upon Denver to hold the most comprehensive discussion of the glue-sniffing problem to date. It was 1967, and far more knowledge, statistical data, and case studies were available than three years prior. In addition, although the assembled group did show the hobby industry's warning film, "The Scent of Danger," it did not include HIAA lobbyists or industry executives on the program.[3] The conference was to be the line in the sand, the map for how to proceed to stop the epidemic in its tracks finally.

It would be a line in the sand in another way, as well, however. Since the 1910s American thinking about the nature of narcotic dependence had been shaped by a mutually dependent relationship between science and law enforcement, governmental agencies seeking to limit access to drugs and dole out punishment for their use, and scientific organizations like the American Medical Association signing on to those witch hunts in order to maintain influence in their respective fields. Historians like Caroline Jean Acker have demonstrated that this incestuous bureaucratic relationship ultimately created the modern image of the deviant—the impoverished urban male criminal. That perception of deviance had just been confirmed the previous month, in December 1966, with yet another in a string of FBI raids on Timothy Leary's Millbrook estate. While the assumption of criminality in glue sniffers was moderated somewhat by the average age of the participants, the deviant model had held.[4]

The January 1967 summit would begin the academic journey across the Rubicon of such conceits, at least as it pertained to model airplane glue. Those who still held fast to traditional notions of deviance and the urban male sample refought the battle they had been fighting since the Harrison Narcotics Act of 1914, but there were new voices that began to question these historical notions. Deviance was a product of the societies that defined it. Criminalization only increased stigmas and furthered what was, ultimately, a self-fulfilling prophecy, leading to more scare headlines and more collaboration between scientific societies and law enforcement officials. It was a new interpretation of the glue sniffer. It would, in the following years, become the dominant academic interpretation of the drug user in general. There, in the cold winter of Denver, the picture of the abuser of model airplane glue began to change.

The summit was hosted by juvenile court judge H. Ted Rubin, from Harrisburg, Pennsylvania. After attending Penn State, Rubin had moved to

Cleveland to become a social worker and earn a master's degree in social service administration from Case Western. Then it was on to Illinois, where he worked in a children's home and earned a law degree from DePaul. In 1956 Rubin moved to Colorado, where he worked at the University of Colorado Medical Center's Children's Psychiatric Clinic, practiced law, and ultimately won election to the Colorado House of Representatives. When Rubin was still in high school in Harrisburg, in 1940, judge Philip B. Gilliam became Denver's only juvenile court justice. The problems with juveniles, however, had only been mounting, and by 1964, the state legislature decided to add an additional judgeship, which Rubin won later that year. He would ultimately serve two terms, his time on the bench lasting until 1970, and he was known primarily for two things: he helped revise the Colorado Children's Code to provide due process for juvenile criminal defendants, and he took the lead on the glue-sniffing epidemic in Denver and Colorado.[5]

When the summit began early on Monday, January 23, 1967, Rubin stood on the dais with John Love, Colorado's governor, and Thomas Currigan, Denver's mayor. They looked forward to progress on the problem and its potential quarantine. They were so glad that everyone had arrived safely. John Scanlon, director of the US Office of Juvenile Delinquency and Youth Development, gave his own introductory remarks on the requirement that any solutions to the glue-sniffing epidemic would need to include both prevention and treatment elements. Then when the introductory remarks had run their course, the leaders welcomed Joseph Lohman, dean of the University of California's School of Criminology, to the stage.

Lohman was a fitting keynote speaker. He was from Denver and had received his AB from the University of Denver before moving on to the University of Chicago. There he immersed himself in Illinois and national politics. Lohman focused often on racial issues, directing the federal government's first major race study through its Commission on Segregation from 1946 to 1948. He worked on Harry Truman's National Planning Commission for Washington, DC, and then worked on Eisenhower's. He also worked extensively in Illinois criminology and corrections, served as sheriff of Illinois's Cook County and state treasurer.[6] After politics, he moved to Berkeley, and now he was back home in Denver, staring out at his waiting peers, desperate for answers to the questions generated by every new glue-sniffing arrest.

Crime, he reminded his listeners, had doubled in the last generation, growing beyond society's ability to police it. But that growth in crime was the result of "a totally new set and complex of social relations."[7] It was, if nothing else, "the lengthened shadow of the community." The melting pot of the United States had given way to greater segmentation into various race, class, and social groups, leading individuals to become alienated, lonely, bored. And alienation, loneliness, and boredom were the principal ingredients fired in the crucible of potential crime. This detachment wasn't just an end in itself. When society was splintered into so many different groups, with so many different understandings of what proper law and morality should be, there was a legitimate disconnect between individual beliefs and enforcement of the law. "It may well be that what we observe as 'disrespect for law' is a normal reaction of normal people to an abnormal condition."[8] And that abnormal condition was, above all, a situation where American institutions had not kept pace with the changing modes of society. It was ultimately a crisis of services.[9]

And crises of services created a "revolt of the clients"—a rejection of the generally assumed tenets of school or the police or welfare or any agency that seemed to represent received rules that such disaffected "clients" had no say in producing. It was, Lohman argued, a "contemporary American revolution" that was driven by three principal factors: rapid population increase, the development of technological advancements, and the tumult of the civil rights movement. Each combined to create a kind of synesthesia, a sensory overload that led to alienation and its antecedents. A growing population and a further segmentation of that population drove the disconnect, as did the multiplicity of new technologies. Technology, in its most basic form, shrinks and expands the country at the same time, and in the process it fundamentally changes the experience of existing in that country. Then, of course, the civil rights movement had challenged the legitimacy of the laws of a bureaucracy that claimed a basis in equality but existed to elevate one race over another. Everyone, every "client," was a victim of such changes, potentially left in the wake of the evolution of the country's state of being, but those feeling the bureaucratic disconnect most were necessarily going to feel the strength of the wake at higher levels than everyone else. "We cannot engage them effectively as individuals," Lohman claimed, "because they are under a collective influence to which they are correspondingly responsive, and which is in competition, even overt conflict, with the remote and formal influences of the school, the police, and other institutions of society." So even though crime could be considered functionally nonconformist, a revolt against such institutions that would-be criminals find confusing or obsolete in a changing world, it was "as much a reflection of the patterning of social life as are the conformist expressions of society." Crime, then, was both nonconformist and conformist, criminals rejecting received group identities in favor of their own.[10] And if anyone was going to reject received group identities in favor of their own, it was going to be children, who seemed both socially and biologically engineered to fall under Lohman's rubric of crime.

More than half of the US population in January 1967 was under twenty-one. More than 60 percent of crimes were committed by defendants under eighteen. Those statistics existed because of misconceptions about America's youth. The marginalization of certain children was a process that could ultimately affect any juvenile, because those marginalized children weren't qualitatively different from their peers. "The society which produces the conformist and conventional middle-class youngster," Lohman warned, "also produces the non-conformist, the deviant or delinquent youngster who is not effectively engaged by the conventional institutions of society." It wasn't that delinquent kids weren't socialized, they were just socialized into other groups and norms at variance with the dominant assumptions of society. But when that society kept fracturing into all kinds of increasingly separated groups, "dominant assumptions" meant less and less. "The point is that social problems are more frequently a conflict between groups rather than a conflict between individuals and a group."[11]

Lohman made the case that prior to the mid-twentieth century, American society was directed toward a single standard, a single group of values. It was, in other words, centripetal. But as of the mid-1960s, "American society today tends to be centrifugal in character; its elements are being forced outward, thus, forming smaller groups, each with its own local pool of experience and culture." Lohman singled out racial minorities, juveniles, and the impoverished as being the principal inheritors of the problem. And the problem was one of definitions. Kids singled out as being somehow different from the mainstream developed an attitude of difference, aligned their thinking of themselves in a way that matched the dominant view of who and what they were.[12]

Meanwhile, the doomsday prophecies centered on the assumption that the new generation of juveniles had become callous, shedding the sense of shame and guilt that previous generations had used to regulate behavior. Lohman disagreed. "Shame reinforces locally defined, non-conforming behavior," he noted.

> In the current cultural communities of America, a variety of subcultures has made shame and guilt problematic. They reflect the local subcultures to which the young people relate. Hence, it produces non-conformity rather than conformity to the established norms of the general society, and correspondingly, it emphasizes attitudes which have local value. The power of the subculture to effect this reversal of values and norms is not fully appreciated.

We are still asking ourselves why it is that formal ordering and forbidding techniques are not able to produce conformity.

Policing was never going to work without some kind of agreed-upon standard of "informal systems of control." And those had disappeared through the fracturing of society.[13]

Lohman compared the plight of the young to the plight of minorities. Relating to the dominant majority (in this case adults) only ended in low self-esteem, leading to a new set of attitudes and values that provided the sense of worth that was unavailable in the adult world. "And more frequently than is good for the society, this turns out to be a condition of defense, a condition of rebellion, rather than a condition of identity with the adult culture." And this meant something. Children of previous generations were integral parts of the family's success, economically or otherwise, and provided a sense of significance. Chores weren't invasions on the freedom of kids. They were "meaningful additions to the family income. The family larder, indeed the security of the family itself, often depended on the contribution of its youthful members." As of the 1960s, however, adults didn't need children like they used to. "For the most part, they are in the way; they are a burden." Such roles, Lohman argued, were easily internalized and dramatically affected the power relations between adults and children. Kids needed "the sense of being needed" but instead found themselves marginalized by the family's lack of requirement for their services.[14]

It was an interesting argument, to be sure. But even those in the crowd who found it hard to believe that the diminished necessity of chores could create an epidemic such as the one faced by the glue-sniffing pundits surely understood more readily the consequences of the fissures in social norms. "It is not that there is a culture of crime," Lohman told his audience. "It is that there is such a plurality of subcultures that the problem of the individual's adjustment to commonly-accepted norms is confounded and that deviance and opposition to law and authority are generated as a matter of course." The answer was to interpret crime not as a problem with law enforcement but, instead, as a breakdown in family structure, job availability, education, and housing. "It is a myth that man's behavior can be changed directly," said Lohman. "It can be changed only by altering the conditions which underlie his behavior."[15] The same could be said for man's younger brother.

The crowd applauded. They gave the keynote address the proper standing ovation and then headed out of the auditorium for lunch, surely discussing Lohman's claims in hushed tones. When they returned, they got down to the more specific business of glue sniffing. Immediately, the speakers ignored everything Lohman had to say. It was the great lesson of the glue-sniffing debate. Theories clashed. Progress only led to more questions. The quixotic nature of the fight—the very nature that the conference was designed to eliminate—appeared before the first day had run its course. William Meloff, an assistant professor of sociology at the University of Alberta, began the post-lunch session by discussing glue sniffers' diversion from the general anomie of national opinion.[16] It was almost as if Lohman's words had already dissipated into the air above the overpriced entrées and early afternoon cocktails.

But Meloff's statements were in aid of a larger study of Denver glue sniffers that sought through comparison with similar non-sniffers to develop a composite of prime candidates for the problem—broken homes, low-income families, low intelligence quotients, poor performance in school. It was common ground, trod by many in the first half of the decade. Meloff and his team, however, went further. They supplied attitude and behavior tests to glue sniffers and control groups, first asking about the level of acceptance for problematic behaviors in three categories (school-related attitudes, law-related attitudes, and individual-related attitudes), then asking about actual participation in such events. School scenarios included smoking on campus, truancy, and cheating; law scenarios included sniffing glue, stealing, and driving without a license; individual scenarios included swearing, disobeying parents, and getting into fights.[17]

The results were instructive. While the glue sniffers had a far different attitude about school-related scenarios, assuming that poor behaviors weren't so bad at all, attitudes about law-related scenarios divided on class lines rather than lines relating to substance abuse. Perhaps even more surprising, there was no appreciable difference between anyone's attitudes toward individual scenarios. The actual participation questionnaire demonstrated a vast differential, with glue sniffers far exceeding everyone else in problematic behaviors, but such was to be expected. So too were glue-sniffer attitudes about school. "It would appear that the school, more than any other factor in the boys' lives contribute to their negative self-concepts and resultant states of anomie."[18] The notion that interpretations of the law divided along class lines rather than actual criminal activity was significant. The assembled could only take away from the study that any boy in a low-income household was susceptible to glue and other potential criminal behaviors. In addition, the fact that there was no difference in attitudes about individual behaviors like parental disobedience and fighting only seemed to indicate that the net needed to be cast even wider. There was no uniformity as to kids' interpretations of what constituted problematic behaviors and attitudes, but all of

## Attributes, Attitudes and Behavior of Glue-Sniffers and Comparison Groups

| A. Background Differences | Glue-Sniffers | Group A | Group B |
|---|---|---|---|
| 1. Average age | 13.5 | 13.8 | 13.2 |
| 2. Ethnicity: White | 0 | 4 | 63 |
|     White, Spanish surname | 34 | 10 | 0 |
|     Negro | 2 | 16 | 1 |
| 3. (a) Living with natural parents | 22% | 44% | 83% |
|     (b) Living with mother | 56% | 25% | 9% |
| 4. Parents divorced | 42% | 19% | 1% |
| 5. Family income: | | | |
|     $1,000 – $2,999 | 31% | — | — |
|     $3,000 – $4,999 | 42% | — | — |
|     $5,000 – $6,999 | 10% | — | — |
| 6. Grade point average last year | 1.79 | 2.16 | 2.8 |
| 7. Days absent from school last year | 60 | — | — |
| 8. Average I.Q. | 86 | — | — |
| 9. Average National Achievement Score | 55 | — | — |
| B. Attitude Differences* | | | |
| 1. School-related attitudes | | .02 | .001 |
| 2. Law-related attitudes | | — | .05 |
| 3. Individual-related attitudes | | — | — |
| 4. Total rejection illegitimate means | | — | .05 |
| C. Behavior Differences* | | | |
| 1. School-related deviance | | .001 | .001 |
| 2. Law-related deviance | | .001 | .001 |
| 3. Individual-related deviance | | .05 | .02 |
| 4. Total conforming behavior | | .001 | .001 |

* The figures given under parts B and C indicate the probability that the differences found in attitudes and behavior between the glue-sniffers and each comparison group could have occurred by chance. Thus a .001 level of significance indicates that the probability of the differences occurring by chance, rather than some other factor, is one in a thousand.

BEHAVIOR AND BELIEF AMONG GLUE SNIFFERS AND NON-GLUE SNIFFERS: William Meloff demonstrated that the behavior differences between sniffers and non-sniffers was significant, while their attitudes about such behavior did not necessarily show such divergences in thinking. Source: William A. Meloff, "Deviant Attitudes and Behavior of Glue-Sniffers in Comparison with Similar and Different Class Peer Groups," in *Presentations of a Conference on Inhalation of Glue Fumes and Other Substance Abuse Practices among Adolescents*, ed. Ted Rubin (Denver: Denver Juvenile Court, 1967), 80-81.

them, whether glue sniffers or not, lower-class or upper-class, saw some of their received behavioral norms as bankrupt in one way or another.[19]

The largest consequence of such results was the realization that, as Robert Hanson explained, taking the reins from Meloff, "glue sniffers are not essentially different from other juvenile delinquents." Glue sniffing was just one form of a broader phenomenon of delinquency. This being the case, "contemporary theories of deviant behavior can thus be expected to provide a range of factors which are pertinent to the explanation of juvenile glue sniffing behavior and related delinquency."[20] There was, among all delinquents, a breakdown in the transmission of social norms (even though Lohman had begun the conference by denigrating the use of such categories, arguing that they actually *created* the functional definition of deviancy and ushered in a self-fulfilling prophecy). Hanson was treading familiar ground. It came as no surprise that he saw family disorganization, poor school performance,

Predictor Variables Associated with Glue Sniffing and Other Delinquency

| Predictor Variables | Criterion Variables | | | |
|---|---|---|---|---|
| | Glue Sniffing | School Deviance | Legal Deviance | Total Deviance |
| 1. Days absent from school last year | .67 | .62 | .63 | .63 |
| 2. Last year's grade point average | .61 | .51 | .59 | .53 |
| 3. Disorganized parental marital relationship | .51 | .44 | .50 | .49 |
| 4. Acceptance of school normal violations | .38 | .54 | .48 | .55 |
| 5. (Lack of) Ego Strength | .34 | .40 | .52 | .49 |
| 6. Acceptance of legal norm violations | .34 | .37 | .46 | .44 |
| 7. (Lack of) Social Dominance | .35 | | | |
| Multiple R | .76 | .76 | .78 | .77 |

USING SEGMENTED VARIABLES TO PREDICT THE POSSIBILITY OF GLUE USE: Robert C. Hanson, using the known statistical data about glue sniffers, gauged various predictors as to the use of model airplane glue and other forms of delinquency. Source: Robert C. Hanson, "Explaining Glue Sniffing and Related Juvenile Delinquency," in *Presentations of a Conference on Inhalation of Glue Fumes and Other Substance Abuse Practices among Adolescents*, ed. Ted Rubin (Denver: Denver Juvenile Court, 1967), 95.

and a tendency toward escapism as the fundamental elements of the glue-sniffing situation. But such only created a potential breeding ground for the would-be sniffer.[21]

"For glue sniffing, the question is: why do some kids become glue sniffers when others in the same family or neighborhood do not?"[22] For Hanson, the answer lay in differential association. Glue entered a neighborhood, usually through a male child who did poorly at school and was from a disorganized, low-income family. Then it spread as more kids with similar backgrounds learned the behavior and used it to seek out companionship. At this point, the child would be labeled a glue sniffer, "then his glue sniffing becomes an expected attribute of one of his social roles."[23] It was a roundabout way of connecting with Lohman's original argument, but Hanson did seem to get there in the end. The idea that children were overly susceptible to the status of definitions was a helpful realization, but for the most part, the ground trodden by Meloff and Hanson was well covered by others who came before them.[24]

Alan K. Done, an associate professor of pediatrics at the University of Utah, followed Hanson. He had the task of discussing the biomedical consequences of model airplane glue and, in particular, placing it within a broader understanding of substance abuse. And the broader understanding was the thing. The fact that glue sniffing had become a national phenomenon—had dominated headlines and convinced many that the sky was falling—tended to create the assumption among many that there was something fundamentally unique about model airplane glue. But there wasn't. Abused substances had three commonalities, no matter what they were. They distorted the mind, they created a degree of habituation, and they created in the body a tolerance to the substance. This habituation could be psychic or physical, so addiction debates were worthless in this context. That they each did different things to the mind was also beside the point. To this end, researchers needed to "consider the constituent parts of the substance abuse problem in proper perspective and in relationship to one another," Done argued. "Any efforts to solve the problem must take into account the overall picture if they are to be effective."[25]

The same could be said for model glue. Emphasis on that one product often blurred the fact that "any fat-soluble, volatile organic solvent is capable of producing similar effects." Done noted that many glue sniffers "graduated" to lacquer thinner because it also contained toluene and could be purchased in larger quantities. Still, toxicity and bodily harm varied depending on the solvent that was being inhaled. Certain chlorinated hydrocarbons could clearly cause long-term organ damage and could even lead to sudden death due to heart-related defects. Toluene didn't produce the same results. Despite the studies that made such claims in the past, toluene, according to Done, did not cause bone marrow damage, and it didn't cause acute anemia. It didn't cause liver damage. It didn't cause leukemia. The only real substantial bodily harm resulting from chronic exposure to toluene was pus and protein in the urine, and Done wasn't even sure that was all so significant. It was theoretically possible that such were the signs of long-term kidney damage, but so too was it possible that the chemical simply had an irritative effect on the kidneys that would dissipate after chronic sniffers stopped sniffing. And while there were clear electroencephalographic abnormalities and an increase in spinal fluid pressure during periods of intoxication, there was no evidence that sniffing glue produced long-term brain damage.[26]

Done was quick to point out that he in no way wanted to minimize the problems posed by the glue-sniffing epidemic. He reminded his audience that glue sniffing was problematic enough without the hyperbole that kept law enforcement officers and researchers chasing false leads. "Also, parents who have solvent-sniffing children have enough to worry about with the true facts, and hardly need be 'scared out of their wits' by other data."[27] Now, it seemed, the glue summit was getting somewhere.

Or maybe not. John William Rawlin followed with a discussion of intravenous amphetamine use as a replacement for heroin when supply seemed to be low. It took the group off topic but still managed to contradict Lohman. Each substance, he argued, needed to be "categorized with regard to the types and levels of its actual and potential abuse capabilities and other relevant sociological, psychological, and medical factors."[28] That was the only way we were going to fully understand the narcotics problem in the United States, whether it be household solvents or high-grade heroin.[29]

Rawlin was followed by Victor Gioscia, the director of research at Jewish Family and Children's Service in New York. His study of glue sniffing focused on Long Island, where the users were middle- and upper-class kids—kids who had no meaningful contact with the low-income families of earlier studies. He introduced his talk by promising to "make some psychoanalytic noises which I love to do because it confounds the anomie theoreticians."[30] If the goal of the glue summit was developing a working consensus, things had already gone far off the rails.

"We are *not* here dealing with an adolescent problem," Gioscia told his audience. "We are dealing with a *pre*-adolescent problem, with kids who are 9, 10 and 11 years old. And that is not adolescence."[31] This being the case, he argued, there was not so much a breakdown of the superego that caused such use but, rather, an immaturity of the superego, a lack of development because of the young age of the user. Never mind that there were myriad examples of adolescents using model airplane glue, and that such use devalued any claims to undeveloped superegos. Gioscia remained steadfast, though his audience was most certainly wary of such claims. He continued by describing the culture of machismo in Latino neighborhoods that could possibly incite such behavior. He also took on claims of addiction, arguing that "the withdrawal syndrome is really an acted out birth trauma," a glorified form of separation anxiety.[32] His argument that "the transition from childhood to adolescence, not the transition from adolescence to adulthood, is the problem that glue is being used as the medication for"[33] had legitimate merit and was an insight that provided something useful in understanding the recourse to glue over and above other substances—availability was less a factor than some had thought; glue was different because the users were different, because the users were dealing with a different set of difficulties unique to the transition into adolescence; when their transitions changed, their "medications" for dealing with those transitions would change, as well.[34] But with that exception, it was difficult to take Gioscia seriously. After a coffee break following Gioscia's talk, the assembled researchers had a discussion period on the afternoon's presentations, and then they adjourned for the day.

The second day began with a presentation by Lieutenant Richard Davis of the New York City Police Department's Youth Investigation Unit, which began with a history of the city's glue-sniffing laws. In 1964 the city approved Section 173.19 of the New York City Health Code, which banned sales of model glue to anyone under eighteen unless it was part of a model-making kit. The second law appeared in September 1965, Section 1747F of the Penal Law, which punished possession with intent to use and possession with intent to distribute. The city's department of health and its youth division studied the problem in depth beginning in 1962 and had initially worked in conjunction with the HIAA, which promised to come up with solutions on its own. The problem, however, kept growing. New York witnessed 780 glue-sniffing cases in 1962, and 2,003 cases in 1964. And so legislation was deemed necessary. Police were powerless to prevent the problem otherwise, and regardless of any significant health effects, the dangers of New York City made glue sniffing a deadly endeavor regardless. Perhaps the most important factor in New York's decision, however, was that "the manufacturers had demonstrated their inability to cope with the problem."[35] It was something that everyone in the audience understood. It was the reason the industry had been relegated to showing a film at the conference, rather than playing any significant role. Despite the continued problems with model airplane glue, the HIAA had never lived up to its promise to successfully change the formula for its toluene and benzene-based compounds.[36]

Davis provided several case studies of lower-class boys between nine and fourteen years of age, each with a problematic home life and some kind of parental neglect. Some became wards of the state. Some received help. One of Davis's examples fell to his death from the roof of his apartment building. But the broader statistics were even more telling. Among sniffers, 89 percent "were also involved in other delinquency."[37] Only 30 percent were from homes with married parents. In one sample of 468 cases, thirty-four subsequently began using other, harder narcotics. "This figure appears relatively low," said Davis, "however, it is significant that a good portion fell between the ages of eight and thirteen years, approximately 60 per cent."[38] The message was clear: the likelihood of cyclical delinquency and chronic drug use increased when the glue sniffers were among Gioscia's "pre-adolescents."

Dale F. Ely, supervisor of attendance for the Unified School District of Long Beach, California, and Milton Luger, director of the New York State Division of Youth, followed Davis. They discussed the role of school administrators and state officials in helping to diagnose trouble spots and to find the juvenile glue sniffers a safe, effective means of being rid of the scourge of

The following are the statistics on glue sniffing within New York City for the period from 1962 to 1966, inclusive:

|  | Male | Female | Total |
| --- | --- | --- | --- |
| 1962 (8 months) | 560 | 32 | 780* |
| 1963 | 1861 | 142 | 2003 |
| 1964 | 1184 | 123 | 1307 |
| 1965 | 1053 | 120 | 1173 |
| 1966 (8 months) | 612 | 96 | 708 |

*Data reflects original source. Note: In September, 1962, a special category was set up for glue sniffing. This accounts for lack of statistics before that date.

INCIDENCES OF REPORTED GLUE SNIFFING IN NEW YORK CITY: Data from Richard Davis of the New York City police department demonstrated that legislation seemed to reduce the incidences of reported glue sniffing in New York. Source: Richard Davis, "Report on the Problem of Glue Sniffing in Children and the Work of the New York City Police Department and Its Youth Investigation Bureau in Combating This Problem," in *Presentations of a Conference on Inhalation of Glue Fumes and Other Substance Abuse Practices among Adolescents*, ed. Ted Rubin (Denver: The Denver Juvenile Court, 1967), 106.

sniffing. Schools needed to work more with psychologists and authorities. Police actions needed to be coordinated with community leaders. Communication and testing were key.[39] Their talks were helpful, and certainly in the grand tradition of do-goodism that the conference was established to promote, but there was nothing in their speeches that left the assembled crowd with anything they hadn't already heard before.

And this was fine. They were warm-ups for the luncheon speaker, New York Medical College psychiatry professor Richard Brotman. "You might expect me to tell you about the disastrous personal consequences of [substance abuse among adolescents], or the terrible social problems it creates. I am not going to do that." Brotman was frustrated. With his profession. With the conference. With the trajectory of the glue-sniffing debate.

> We all keep saying pretty much the same thing: we used to emphasize that there are different kinds of drug *users* and that differential diagnosis ought to be introduced into their treatment; now we try to get people to understand that there are different kinds of drug *use*, too, and that one of the kinds is conformist adolescent use. The main result of all this educational effort is that we get a plethora of dreadful legislation and an hysterical press.[40]

Traditionally, psychiatrists treated the affluent, or they were paid by governmental agencies to treat the poor, who were often unable to communicate effectively or simply didn't have the time for such psychiatric care to work. At the same time, social workers "have also traditionally been paid by one group to render service to another." But that system was changing, pushed by the fact that "the poor are beginning to find themselves in a better position to insist upon more meaningful and appropriate mental health care" and the realization by middle- and upper-class parents that their own children were abusing substances. In the long term, such developments were all for the betterment of better psychiatric care, but in the short term, adolescents were going to get the short end of the stick.[41]

The problem, argued Brotman, was with the massive number of students who recreationally used substances like model airplane glue or marijuana and who were in no other way delinquent. Drug abuse in any form was not a crime directed at others, and crimes without victims were always problematic. So maybe it was a problem because it was a threat to the existing social order instead. If so, then what exactly was the nature of the threat? It was a good question. Brotman saw substance abuse as one aspect of a broader objection to the condition of contemporary society, pushed by the smaller society of a group of peers infatuated with the substance. He cited the FBI's raid on Timothy Leary's Millbrook, which had occurred earlier that month, and the peaceful reaction of those arrested. It was the criminalizing laws that created violence, not the drugs. "The funny thing about this is that the kids have more factual knowledge about the drugs than their parents do. All the adults can think of is to forbid it."[42]

He closed with a bang. He reminded everyone assembled that certain forms of substance abuse prevalent among middle- and upper-class teenagers was not associated with any psychiatric disorder. "Drug use is but one facet not of rebellion, but of dissent; not of social irresponsibility and immorality, but of adherence to a set of rational and humane tenets, quite like those upon which, in theory, our democracy is founded, but which are not infrequently rather contrary to the way we actually do business."[43] There was, in Brotman's hands, a certain grace in drug use, at least in the use of nonaddictive drugs like glue or marijuana or LSD. There was no way to eliminate it through legislation, and frankly, legislation was causing the confusion and undue hysteria that plagued the headlines of American newspapers.[44]

It was a virtuoso performance and provided a counterweight to everything else uttered at the summit. Maybe there was no psychopathology. Maybe there was no need for legislation. The audience surely listened with furrowed brows.

Ted Alex, a group therapy consultant for the Denver juvenile court's glue-sniffing project, restored their confidence in themselves. He gave a presentation describing the Denver Glue-Sniffing Project, which focused on two groups of kids, one from a predominantly Latino neighborhood where the boys all acted as a group and readily admitted to sniffing glue, the other from a middle-class white neighborhood where the boys were reluctant to admit use because of a stigma of low status attached to the practice. Workers in charge of each group emphasized individual counseling combined with family therapy and work in the community. Neighborhood counselors were available twenty-four hours a day. Special classes allowed the boys to work at their own pace. Full-time group workers stayed in the neighborhoods. It was a valiant and successful effort, but the conclusions about glue sniffers themselves seemed like common fare. Glue sniffing was a socioeconomic problem. The kids lived with high levels of anomie and glue provided a mechanism of escape. The full report on their twenty-four-month study would appear in June.[45]

The day's final presentation was from Robert Sobolevitch, assistant superintendent of South Philadelphia's Youth Development Center. Sobolevitch took the focus off of Denver and New York, the two metropolitan areas most obsessively and consistently studied. Only sixty-five instances of glue-sniffing arrests occurred in Philadelphia in 1965 (three-tenths of 1 percent of juvenile cases in the city). Philadelphia had no glue-sniffing statute like those in New York and Denver, and it would seem that such statistics argued that the city had no glue-sniffing problem. "In fact, nothing could be further from the truth."[46] Although the Philadelphia city council was debating a glue-sniffing law, the fact that it didn't have one understandably skewed the arrest figures. Welfare and social workers had a far more substantial knowledge of the reality of glue sniffing in the city. The problem existed most readily in middle-class neighborhoods. In the northeast part of the city, the neighborhoods most affected were Jewish and Catholic, one area in the heart of the Jewish community even known as "Sniffers Nob."[47]

The problem for Philadelphia, however, was that with so many other issues (violent crime, one-fourth of its population living in dangerous slums, etc.), comprehensive glue-sniffing support programs fell by the wayside. This left the problem to nongovernmental groups like the Jewish Family Service to work with juvenile glue sniffers. Still, what Sobolevitch found most interesting was that, although glue sniffing was growing and presaged an increase in hard-core drug use, Philadelphia didn't have the kind of narcotics problem of New York, just ninety miles to its north. The city was working to study the problem further, but Sobolevitch couldn't resist adding his own psychoanalytic analysis of glue sniffing. "Although the individual glue sniffer may be disturbed, he is reacting to certain pressures which are placed upon the entire group and . . . this behavior is constantly being reinforced within the group." This being the case, it was impossible to consider the glue sniffer an "individual deviant from his own society."[48] His own society, after all, was reinforcing the validity of his actions.

It wasn't exactly a stunning new revelation, but Sobolevitch did have the benefit of being right. And the emphasis on an area that demonstrated clear differences with the glue-sniffing hubs of Denver and New York was valuable. The audience received the speech warmly and then had coffee before an open discussion. The second day of the summit closed with a thirty-minute film produced by the HIAA, but representatives of the HIAA weren't invited and didn't profess any of their typical promises either before or after the screening. Everyone looked on politely, appreciating the effort but ever mistrusting the association's sincerity. Day Two came to a close.

The final day of the conference was consumed with workshops, where participants banded together in groups to discuss their own experiences with glue sniffing and their hopes and plans for further prevention.[49] They were followed by an adjournment by Ted Rubin, lead architect of the glue summit. "I trust it is clear that if these youngsters are to be helped, the entire spectrum of community agencies and child-serving resources must be heavily involved. This includes the police, the schools, community centers, mental health agencies, and the courts."[50] Legislation was valuable, but whatever legislation municipalities and states created had to be careful not to label children infected with the glue-sniffing disease as delinquents. Such children required help, not punishment.[51]

Schools needed to be vigilant about behavioral problems that hinted at an abiding glue-sniffing menace. Community centers did, too. Caseworkers needed to focus on each element of a glue sniffer's family, trying to heal relations throughout the sniffer's life. Mental health workers needed to give of their time at clinics that reached lower-income children without the resources to make appointments. Police officers should use judgment that deemed nonviolent offenders as causes for concern rather than arrest. "In these instances, to the degree feasible, take the child to his home." Rubin's final plea was to his fellow juvenile court judges. "An habituated glue-sniffer will not stop this practice because you warn him or threaten him," he scolded. "Nor will he attend school daily though you warn him of the consequences. Traditional probation and traditional delinquency restitutions do not reach him. Far more total and prolonged intervention is needed. Arrange it."[52]

With that, the glue summit of January 1967 came to a close, and its attendees filed out into the cold wintery slush of downtown Denver. Surely many of them were fired by the conference to enter their communities and rededicate themselves to helping solve the country's glue-sniffing epidemic. The problem, however, was that for all the rallying 'round the flag, most of the participants were still treading on common ground. Joseph Lohman and Richard Brotman had presented new ideas, however, arguing that the traditional way of thinking about the adolescent substance abuse problem wasn't working. Every other presenter who filled the gaps between them continued talking about the outbreak in those traditional modes and with the same traditional tropes, but the model that would eventually dominate the thinking of researchers through the decade's turn was there, fighting for time and space. Millbrook was under heavy surveillance, a broader drug culture was thriving throughout the country, and the research that dealt with it was finally trickling down to those who treated the children who might not be so deviant after all. And so the participants were left with almost as much uncertainty as when they arrived. For all the talk and debate, they could only wait to see what happened next.

# The Pursuit of the Monster

"A MAN WHO COMMITS a crime while drunk is nonetheless responsible for his act," *Time* magazine explained to its readers. "But what of those who commit crimes while under the influence of something more unorthodox?" The magazine told the story of a fifteen-year-old Detroit boy who sexually assaulted and then strangled eight-year-old Deborah Crowther and six-year-old Kimberly Crowther. He had nabbed the girls while they were walking in a field near their home, but prior to the attack, the boy and two of his friends had sniffed fifteen tubes of model airplane glue. And this put juvenile court judge James Lincoln in a difficult position. There was no question that the boy had committed the rapes and murders, but there was a question as to whether he was in his right mind at the time. It was August 1967. The glue-sniffing epidemic had been spreading for eight years, and scare headlines across the country had warned that kids who sniffed glue lost their minds. Lincoln reluctantly decided that the boy had lost his mind. "The boy is not guilty of the charge by reason that he was incapable of controlling his actions at the time of the killings." Instead of a prison sentence, the boy would stay in a mental hospital or training school until his nineteenth birthday. It was an outrage to the community. It was an outrage to *Time* magazine. And it prompted even more scare headlines like the ones that influenced the judge's decision. But Lincoln felt that his hands were tied. In his ruling, the judge argued for action by the Michigan legislature that would make future similar cases "indefensible on grounds of temporary insanity."[1]

It seemed incomprehensible. Glue sniffers weren't just killing themselves anymore. They were raping and killing others, and they were getting away with it. Frightening, sensational stories like those emanating from the Detroit juvenile court kept model airplane glue on minds and tongues across the nation, and they kept researchers scrambling to come up with substantive answers to the increasingly dangerous dilemma.

Of course, some of those substantive answers were just as frightening and sensational as the news headlines. In March 1967, *The Lancet* published a letter to the editor from toxicologists and pathologists from Pittsburgh's Allegheny County coroner's office. The letter told the story of a white middle-class boy, sixteen years old, found dead in his basement after huffing benzene-laden rubber cement. The boy suffered from acute bilateral pulmonary edema, cerebral edema, liver congestion, spleen congestion, and mucosal surfaces on his stomach and duodenum. Toxicology reports indicated that the infected areas were loaded with benzene. His blood contained ninety-four micrograms of benzene per one hundred milliliters. His kidneys had fifty-five milligrams per one hundred grams. The benzene had spread quickly, a contagion that moved through the body and affected almost every area. The toxicologists were quick to point out that the rubber cement was fundamentally different from most model airplane glues, as the toluene that littered such products was methylbenzene, which existed in much lower concentrations than the pure benzene in rubber cement.[2] There was no time for the Pittsburgh boy to seek out victims in a local field. The benzene overtook him before he ever had a chance to overtake anyone else.

The vast majority of glue sniffers neither died nor committed murder, however. In May 1967, the *British Medical Journal* reported the case of an adult heroin user who had begun his experience with narcotics as a glue sniffer in 1962. After hospitalization and treatment, he was discharged to a London hostel, where he graduated to marijuana. From that followed experimentation with heroin, and from that followed addiction. But his taste for model airplane glue never subsided. He used glue as a substitute when he couldn't afford heroin, until March 1967, when he decided to combine the two. He shot up as usual, then began sniffing glue while under the influence. It wasn't long before he lost consciousness. The doctors at St. Thomas's Hospital were able to save him, but after his release he almost immediately began sniffing glue again, was picked up by police for public intoxication, and then returned to the hospital.[3]

The case was significant for a couple of reasons. First, it demonstrated signs of legitimate dependence. Glue seemed just as palpable a force in the man's life as heroin. It was always possible that the man moved to glue upon release because it was his only possible and available substitute for heroin, but the reversion to glue demonstrated, if not chemical addiction, then at least a psychic dependence. Second, it demonstrated that glue was, or at the very least, could be a legitimate gateway to more problematic substances. "Where at one time the typical teenage heroin addict had a preceding experience of experimentation with marihuana," declared Julius Merry at St. Thomas's Hospital, "over the past few years experimentation with marihuana has largely been replaced by other substances, especially amphetamines, barbiturates, cough syrups, and glue." Finally, it demonstrated such conclusions using one representative example. While the possibility of dependence and graduated drug use certainly existed, using only one example to demonstrate these possibilities created an assumption of likelihood that simply wasn't there.[4]

Whatever it might lead to, model airplane glue was a significant problem in its own right. As promised at the glue summit, Lester G. Thomas, director of research and programming for the Denver juvenile court, published the findings of the Denver Glue-Sniffing Project in the summer edition of the 1967 *Juvenile Court Judges Journal*. The report quoted heavily from the January glue summit and provided a predictably muddled portrait of the glue sniffer.[5]

Areas that assumed that glue sniffing was only a minor distraction, for example, soon learned from truants caught in the act that the problem in their area was far more widespread. Model airplane glue was one of literally hundreds (if not thousands) of toluene-based products that kids could find in stores or around the house. If that wasn't bad enough, substance-abusing kids usually turned to other substances as well, such as gasoline, nutmeg, or nasal inhalants, before turning to barbiturates, hallucinogenics, or pot. As Thomas put it, glue sniffers usually suffered from "some psycho-social disorder or dysfunction," but at the same time, somehow, glue sniffers were not "seriously emotionally disturbed, but are merely engaging in activities which dissent from 'socially acceptable' practices." Part of the seemingly contradictory nature of such conclusions stemmed from a crisis of cooks in the kitchen, with "criminologists, psychologists, medical men, sociologists, court personnel, and social workers," all with their own interpretations of the problem. Thomas ignored the fact that each group seemed to be talking past one another. He instead decided that the different arguments from the different disciplines helped to shape the portrait of the glue sniffer and therefore benefit any potential contact trace map.[6]

Still, Thomas admitted that "it is virtually impossible, at this time at least, to construct a 'profile' of a substance abusing child." It was an odd admission given the eight years of study that researchers had devoted to the topic and given that the Denver juvenile court had spent the last twenty-four months intensively working to create such a profile. The admission, however, didn't stop Thomas from giving it a try. Glue sniffers "acted in," as opposed to acting out. They tended to only do harm to themselves, not others (a conclusion surely disputed by the Crowther family). "Psychologically, this type of person was not well put together," but the vast majority of glue sniffers "are neither mentally ill, nor very neurotic." There was usually a poor school record and a broken home. Most of those homes were "economically marginal," but in the same breath Thomas noted that most had an "adequate income." Glue sniffers were ethnic or racial minorities, though black adolescents seemed far less affected than most and the majority of cases throughout the country demonstrated substance abuse by white kids. The children were mistrustful of authority, introverted, suspicious, and yet they were generally "courteous, polite, [and] most cooperative." But this was just a smoke screen, an attempt to get people to leave them alone. Glue sniffers were loners who were dramatically influenced by their peer group.[7]

Every single element of the profile seemed a contradiction. It was as if the portrait of the glue sniffer were an optical illusion, with the kid looking a certain way from one angle, then a completely different way when the picture or the viewer's eyes moved. Thomas seemed to be moving a three-dimensional image back and forth, watching as the glue sniffer appeared one way, then another, one way, then another, one way, then another. It was the very problem that Joseph Lohman had warned against at the 1967 glue summit. It was a vestige of a way of thinking about narcotics that historians such as David Courtwright and Caroline Jean Acker have described as passing out of favor by the late 1960s, particularly when it came to harder, more traditional drugs. And yet the mirage persisted.[8]

Alan Done, another of the glue summit's participants, thought he could explain the contradictory nature of the portrait. In December 1962, the HIAA had polled 337 police chiefs in cities with more than fifty thousand residents about the nature of their glue-sniffing problem. Of the 136 who responded, 67 percent declared that there was no problem in their city; 18 percent admitted there was a problem but stated that it wasn't serious. Only 15 percent understood glue sniffing to be a legitimate local crisis. The disconnection existed because most who sniffed glue never fell under the auspices of juvenile officials. Only those whose use led to some kind of truancy or disorderly conduct ever fell within the purview of the law.[9]

Perhaps a better way to understand the scope of the problem was to investigate which cities and states had passed legislation to deal with the

## Cities and States in the United States That Have Enacted Legislation in an Effort to Control Solvent Installation

| City or State | Use for Intoxication Prohibited | Sale to Minors Restricted | Register Required |
|---|---|---|---|
| Anaheim, Calif. | Yes | No | No |
| Azuza, Calif. | Yes | No | No |
| Baldwin Park, Calif. | Yes | No | No |
| Berkley, Mich. | Yes | Yes | Yes |
| California | Yes | Yes | No |
| Chicago, Ill. | Yes | Yes | No |
| Cincinnati, Ohio | Yes | Yes | No |
| Columbus, Ohio | Yes | No | No |
| Costa Mesa, Calif. | Yes | No | No |
| Detroit, Mich. | Yes | Yes (*M) | Yes |
| Hazel Park, Mich. | Yes | Yes | No |
| Houston, Texas | Yes | Yes | No |
| Illinois | Yes | Yes | No |
| Lakewood, Ohio | Yes | No | No |
| Maryland | Yes | No | No |
| Minneapolis, Minn. | Yes | Yes (*M) | Yes |
| Montebello. Calif. | Yes | No | No |
| Newark, N.J. | Yes | Yes (*M) | Yes |
| New Jersey | Yes | Yes | No |
| New Rochelle, N.Y. | Yes | Yes | Yes |
| New York City | No | Yes (*M) | No |
| New York (state) | Yes | No | No |
| Parma, Ohio | Yes | No | No |
| Providence, R.I. | No | Yes | No |
| Rhode Island | No | Yes | No |
| Roseville, Calif. | Yes | No | No |
| Salt Lake City, Utah | Yes | No | No |
| San Antonio, Texas | Yes | Yes (*M) | No |
| Washington, D.C. | Yes | Yes | Yes |
| Waukegan, Ill. | Yes | Yes | Yes |
| Woodbridge, N.J. | Yes | Yes | Yes |
| Yolo County, Calif. | Yes | No | No |

Total: 8 states; 26 cities or counties.

• Model kit assemblies with glue included is permitted

GLUE-SNIFFING LEGISLATION: Cities and states in the United States that enacted some kind of glue-sniffing legislation by 1966. Source: Edward Press and Alan K. Done, "Solvent Sniffing: Physiologic Effects and Community Control Measures for Intoxication from the Intentional Inhalation of Organic Solvents, I," *Pediatrics* 39 (March 1967): 452.

menace. Done tabulated "most of the states and cities in the United States" that adopted some kind of glue-sniffing ordinance. You don't pass a law, his argument went, without a problem existing. And, of course, even this didn't really demonstrate the extent of the problem because many states and municipalities "deal with it through judicious enforcement of existing legislation against drunk and disorderly conduct."[10]

It was a different and seemingly more effective strategy for understanding the extent of the glue-sniffing phenomenon. Done worried less about an accurate portrait of the sniffer and instead focused on the broader picture of the extent of sniffing. He compared juvenile statistics for New York, Chicago, and Los Angeles. Glue sniffing remained a large portion of the juvenile substance abuse problem in all cities (as compared to alcohol and narcotic use), but in New York, the number of glue sniffers seemed to be on the decline from 1963 to 1965. He ascribed the decrease to legislation,

### Incidences of Solvent Sniffing among Juveniles in New York City, Chicago, and Los Angeles in 1963, 1964, and 1965 in Comparison to Other Forms of "Addiction" and to Total Juvenile Offenses Reported

| Offenses | New York City | | | Chicago | | | Los Angeles | | |
|---|---|---|---|---|---|---|---|---|---|
| | 1963 | 1964 | 1965 | 1963 | 1964 | 1965 | 1963 | 1964 | 1965 |
| "Glue sniffing" | 2,003* | 1,307* | 1,173 | 232 | 476 | 389 | 442 (522 in 1962) | 623 | 594 |
| Alcohol abuse | 811‡ | — | | 878 | 818 | | 693† | 589 | |
| Narcotic offenses | 1,403* | — | | 8 | 16 | | 360 | 760 | |
| Total juvenile offenses | 40,237* | — | — | 37,438 | 40,352 | — | 24,663 | 24,388 | — |

\* These New York City Figure include persons under 21 years of age—1,788 of the 2,003 were under 16 years. Figures for Chicago and Los Angeles are for persons under 18 years of age.

† Drunkenness only, excludes 886 minors in possession of liquor, 29 minors for purchase of liquor, and 54 minors for driving drunk.

‡ Under 18 years of age—use of alcoholic beverages legal at age 18 in New York.

— Breakdown not available as of this date.

GLUE SNIFFING IN THREE CITIES: Edward Press and Alan Done calculated the rates of change in glue-sniffing numbers for New York City, Chicago, and Los Angeles. Source: Edward Press and Alan K. Done, "Solvent Sniffing: Physiologic Effects and Community Control Measures for Intoxication from the Intentional Inhalation of Organic Solvents, I," *Pediatrics* 39 (March 1967): 453.

educational programs, and the state of glue sniffing as a fad that had begun to run its course over the middle of the decade. When solvents were taken as a whole, toluene-based substances like glue were by far the most popular, followed by benzene-laden gasoline, then acetone products such as rubber cement and fingernail polish remover.[11]

Done's research was supplemented by his coauthor, Edward Press, from the Illinois Department of Public Health, who was particularly important in culling statistics from Chicago. Like everyone else, the team followed its portrait of glue sniffing with a portrait of glue sniffers themselves, covering the same ground as those who came before them. Males between ten and fifteen. Latinos and whites, though incidences of black use were on the rise. Poor intelligence quotients and school performance. In some areas, like California, the boys seemed to be from large, poor families. In others, like Salt Lake City, the boys were from smaller families in every income bracket. In both areas and in most others, those homes were usually broken, without a dominant paternal presence. "There is insufficient data to support or refute the possibility that sniffing is causally related to the later adoption of

### Age and Sex of Juvenile Solvent Sniffers Processed by the Youth Division of New York, Chicago, and Los Angeles Police Departments 1963–1965

| Age | New York City | | | | | | Chicago | | | | | | Los Angeles[†] | | | | | |
| --- | --- | --- | --- | --- | --- | --- | --- | --- | --- | --- | --- | --- | --- | --- | --- | --- | --- | --- |
| | Male | | | Female | | | Male | | | Female | | | Male | | | Female | | |
| | '63 | '64 | '65 | '63 | '64 | '65 | '63 | '64 | '65 | '63 | '64 | '65 | '63 | '64 | '65 | '63 | '64 | '65 |
| 7–9 | 49 | 21 | 11 | 1 | 1 | 0 | | 22 | 3 | | 1 | 0 | | | | | | |
| 10–15 | 1,612 | 967 | 87[1] | 126 | 101 | 100 | | 355 | 315 | | 34 | 15 | | | | | | |
| 16–20* | 200 | 196 | 171 | 15 | 21 | 20 | | 60* | 71 | | 4* | 3 | | | | | | |
| Total | 1,861 | 1,184 | 1,053 | 142 | 123 | 120 | — | 437 | 389 | — | 39 | 18 | 418 | 586 | 529 | 24 | 13 | 65 |

*16 through 17 years for Chicago; 16 through 20 for New York City
† Breakdown for Los Angeles not available
1963 breakdown for Chicago not available

CITY GLUE-SNIFFING DIFFERENTIATION BY AGE AND SEX: Press and Done's research on New York, Chicago, and Los Angeles provided relatively full age and sex data for New York, some for Chicago, and none for Los Angeles. Source: Edward Press and Alan K. Done, "Solvent Sniffing: Physiologic Effects and Community Control Measures for Intoxication from the Intentional Inhalation of Organic Solvents, I," *Pediatrics* 39 (March 1967): 455.

more seriously deviant behavior,"[12] they wrote, but some kind of relationship seemed more than likely. While sniffing glue, users were likely to be prone to harmful accidents, but long-term damage to organs such as the liver and kidneys was still uncertain.[13]

This wasn't going to stop the researchers from trying to draw conclusions anyway. A Swedish glue-sniffing study had demonstrated hepatic enlargement of the liver. Other studies showed "dilated pupils, nystagmus, diminished reflexes, and muscular tremors."[14] Such were combined with immediate attacks on the central nervous system, the phenomenon that provoked the high provided by model glue. Though some sniffers reported substantial hallucinations, it was gasoline, nutmeg, and other household products that provided hallucinogenic effects comparable to psilocybin, mescaline, or LSD. A growing tolerance and clear habituation existed, even though Press and Done—along with everyone else—had a problem with loaded terms such as "addiction." Hematologic studies demonstrated that many glue-sniffing patients, though by no means all of them, showed low hemoglobin levels. There was basophilic stippling, poikilocytosis, anisocytosis, and hypochromia. Urinary studies demonstrated pyuria, hematuria, and proteinuria. Renal and liver function tended to be relatively normal. There were no incidents of permanent brain damage from the use of model airplane glue, though temporary electroencephalographic abnormalities were far from unheard of. Fatalities that didn't result from accidents like falling from roofs or jumping in front of cars tended to result from suffocation after users sniffed glue from plastic bags.[15]

It was a helpful report on the medical effects of glue sniffing that sought to wade through the miasma of hyperbolic claims and find some measure of verifiable proof. But what could be done? "The best results have been achieved with supportive psychotherapy which has been individualized on the basis of the problems which exist in the psychological make-up of the individual or in his environment (usually his family life)," wrote Press and Done.[16] But this was a slow process and didn't provide any immediate relief. Then there was institutionalization, which would force abstinence in a way that psychotherapy couldn't. But "forced abstinence alone is of no lasting value." Solutions needed to be therapeutic, not punitive. "Community-wide mental health measures" could provide a semblance of prevention by solving the kinds of problems that might drive a potential glue sniffer to a tube and paper bag.[17] Legislation could be helpful in making it more difficult for users to purchase glue and could provide a deterrent to those contemplating use. The problem with the laws that had been passed by states and municipalities through the first half of 1967 was that they emphasized toluene-based model

## Laboratory Abnormalities among Toluene "Sniffers"

| Finding | Present Study | Reference 4 | 18 | 1 | 5 |
|---|---|---|---|---|---|
| | Number of Patients | | | | |
| | 16 | 89 | 27 | 15 | 32 |
| **Hematologic** | | | | | |
| Anemia | 0/13 | 20/89 | 0/27 | 0/14 | 0/32 |
| Neutropenia | 0 | 0 | 0 | 0/15 | 1 |
| Lymphopenia | 0 | 4 | 0 | 0 | 0 |
| Eosinophilia (>5%) | 4 | 25 | 2 | 0 | ? |
| **Renal** | | | | | |
| Pyuria | 6/16 | 32/89 | 0/27 | 0/15 | All 32 urines "normal": details not given |
| Hematuria | 3/16 | 14 | 2 | 0 | |
| Proteinuria | 5/13 Urea | 12 | 0 | 1 PSP* | |
| Clearances | 1/7 | | | 0/13 | |
| Azotemia | 0/9 | | 0 | 0/7 | |
| **Hepatic** | | | | | |
| SGOT of SGPT† | 0/14 | 0/89 | 0/27 | 0/5 | |
| BSP retention‡ | 0/12 | | | 0/6 | 2/32 |
| Hyperbilirubinemia | 0/4 | | | 0/14 | |
| Hypealbuminemia | 0/2 | 0 | | 0/12 | 0 |
| Thymol turbdity | 0/2 | 0 | | | 0 |

Toluene source was plastic cement in all cases except for one in the present study and all reference 1 who used paint thinner.

Studies since have been extended to include more than 650 cases and the findings substantiate those shown above; in addition, borderline elevations of serum lactic dehydrogenase have been found in some subjects (unpublished data).

* Phenosulfonphthalein clearance in 2 hours
† Serum glutamic oxelacetic or pyruvic transaminase
‡ Bromsulphalein retention in 45 minutes

ORGAN DAMAGE RESULTING FROM SNIFFING GLUE: Edward Press and Alan Done studied the rate of abnormalities in the blood and organs of glue sniffers. Source: Edward Press and Alan K. Done, "Solvent Sniffing: Physiologic Effects and Community Control Measures for Intoxication from the Intentional Inhalation of Organic Solvents, II," *Pediatrics* 39 (April 1967): 612.

airplane glues. But there were myriad household products and organic solvents that could intoxicate America's youth. "Failure to take this fact into account is not only discriminatory and ineffective, but it also introduces the real danger that habitués will switch from relatively innocuous toluene-containing plastic cements to more toxic substances such as may be found in cleaning and degreasing compounds."[18] It could also send the sale of model airplane glue underground, creating a black market that would make the sale of glue look much like the sale of more dangerous narcotics like cocaine or heroin. Finally, failure to include other organic solvents placed an unfair stigma on model-making and other hobbies, which when engaged in properly provided a healthy and constructive outlet for adolescents.[19]

The legislation had varying levels of success and failure, with some police and legal officials using the laws to make arrests and distinguish glue crimes from those of alcohol on one hand and heavier narcotics on the other. Other officials seemed frustrated, arguing that the measures were impractical and difficult to enforce. As to users, glue laws gave police officers one more misdemeanor to chase. As to sellers, no police force in the nation had the manpower to critically examine every hobby shop in a municipality and rigorously check their records of glue sales to discover any fraud or inaccurate record-keeping. And even if they did have such manpower, no police force had the time to requisition hobby shop sales records and compare them to invoices, somehow magically teasing out which tubes went to adults and which to children.[20] It was an impossible task, a fool's errand that no force could complete anyway.

Education was fraught with similar difficulties. Making community leaders, teachers, and parents aware of the dangers associated with sniffing model airplane glue was an invaluable tool, but it was a task only accomplished by publicity. And publicity related to glue sniffing was filled with hyperbolic scare claims that did less to educate and more to frighten. In addition, publicity was also a problem because, in warning against the potential dangers of model airplane glue and explaining its effects, it gave children who might not otherwise have thought to sniff the substance that came with their models an idea for new recreations. It was, in all ways, the contagion that spread the epidemic. Still, there was no substitute for making communities aware of problems that might affect them, and in areas where glue sniffing had already presented a problem, worries about giving adolescents ideas were really beside the point.[21]

As the glue-sniffing menace continued to develop around concerned researchers and law enforcement officers, however, it was becoming clear that adolescents weren't the only ones getting ideas. In October 1967, two psychiatrists from the University of North Carolina studied two freshman students at Chapel Hill, glue sniffers who seemed to break the traditional perception that users of model airplane glue were the kinds of kids who never made it to college. One of them had been sniffing glue for two years, usually when depressed. He showed no evidence of brain damage, organ problems, or blood dyscrasis. His description of the palliative effect of model glue was instructive. He described a three-stage process that fundamentally affected the kind of high he received. Sniffing a small dose of glue would "produce euphoria, moderate confusion, 'pleasant dizziness,' and tinnitus." A moderate dose would create a dreamlike state, a feeling of disconnection from the self and a state of unreality. Sniffing a heavy dose would create hallucinations similar to psychotropics like mescaline or LSD.[22]

The other sniffer was far more cavalier. He didn't sniff glue to alleviate depression. He wanted a thrill, a sensation. That being the case, he used it more often than his collegiate counterpart. Still, he also showed "no significant physical abnormalities."[23]

None of which was to say that toluene-based substances weren't dangerous. They irritated mucous membranes, attacked the central nervous system, and could cause anorexia, ataxia, and muscle weakness, the psychiatrists claimed. But like others before them, the Carolina researchers warned that toluene-based substances weren't the only ones available. "A brief survey of the contents of the glue section of a hobby store will demonstrate that a variety of intoxicating solvents are available." There were aliphatic hydrocarbons (or hexane), ketones like acetone or methyl ethyl ketone, esters such as butyl and ethyl acetates. There were the aromatic hydrocarbons like toluene and xylene along with butyl and isopropyl alcohols. And then there was benzene. All of them were dangerous, and benzene and acetone could cause significant permanent damage. And so the researchers used two case studies that demonstrated no significant damage to glue sniffers in order to argue for the inherent possibility of significant damage to glue sniffers. It was the problem at the heart of the epidemic. A phenomenon that presented real danger was exacerbated and spread by unnecessary claims of even greater danger.[24]

And dire warnings promised that such danger could take place anywhere. As of early 1968, the same solvent problem that "has reached epidemic proportions in the U.S.A."[25] had reached north into Canada. Canadian federal authorities, having heard the alarm sounded in the United States for the better part of a decade, were "now moving to impose restrictions on the sale and use of model airplane glue and other substances containing volatile solvents."[26] The thrust of the Canadian outbreak, however, seemed to center on Manitoba's capital city, Winnipeg.[27]

Winnipeg poison control had been waiting since 1960 for the epidemic to creep across the border. But poison control never got a call in 1960—or 1961, or 1962. "In 1964 we had our first indication that the practice did exist in the city," wrote Vera Gellman, the director of Winnipeg poison control, when a local pharmacist noticed that there had been a run on one particular brand of fingernail polish remover. The polish remover problem was a real one, and poison control responded quickly by getting the word out across Manitoba, urging pharmacists to take the polish remover from open display, and the problem seemed to abate. "No further incident came to our attention for two years."[28]

In the autumn of 1966, the fingernail polish remover fad morphed into a glue-sniffing fad, and the hysteria that followed led to dire predictions of the end of the universe for Winnipeg schoolchildren. Poison control, however, was content that the vast majority of glue-sniffing cases in Winnipeg were kids who were "just experimenting and hoping to gain the acceptance of their peer groups." It wasn't the chronic problem that it was in Denver or New York. And so the organization's best defense was to counter the scare headlines with education programs for nurses, teachers, social workers, and ministers, those whose work brought them into contact with potential glue sniffers. And there, at least, it seemed to work. Though teachers and social workers continued to fret over the possibility that discussion of glue sniffing would only give children ideas, the problem seemed to abate.[29]

The United States, however, was not Canada, and the spread of the "disease" had moved beyond poison control's ability to quarantine it. At the onset of 1968, twelve states had enacted glue-sniffing legislation (see chapter 8).[30] So, too, had Puerto Rico. In addition, no fewer than twenty-nine counties and municipalities had produced regulatory statutes. The effort was there, if not the results. And the effort would need to continue, if not at the legislative level then at least with further research and information. "It appears that the practice of glue sniffing," wrote Lenore R. Kupperstein and Ralph M. Susman, "as with other 'social' practices, has evolved into a ritualistic pattern with differing degrees of refinement and variation, the adoption and adherence to which are intended to demonstrate some degree of sophistication on the part of the user." It was a frightening thought. The glue sniffers were supposed to be stupid. That was, it seemed, every official's ace in the hole. But as the problem grew, so too did the profile of the user. Experiencing glue as an individual act that often manifested itself in a more social setting wasn't necessarily a sign of psychological disorder.[31] That's exactly what book clubs did. It's what every adolescent did in a variety of different settings—experiencing something individually and often meeting with a peer group to experience that thing together. What was homework if not a version of that?

Kupperstein was the associate director of the University of Puerto Rico's criminology program. Susman was deputy director of the US Department of Health, Education, and Welfare's Office of Juvenile Delinquency and Youth Development. Their principal concern was that the reactions to glue sniffing as constituted in early 1968 resulted in "a social-psychological problem actually within the purview of public health and social welfare" and so had been transferred to courts and correctional agencies, adding unnecessary burdens to those groups while not actually utilizing the best resources available to combat the problem. The relative absence of social welfare and public health agencies in the treatment (or correction, or punishment, or whatever) of glue sniffing left two substantial problems: "(1) an overburdening of the juvenile courts and their resources; and (2) a reduction in the competency and scope of activities of these agencies resulting in a redundant service pattern and a rather conventional and narrow definition of role and function." Leaders were spinning their wheels, placing the onus on juvenile courts—bodies designed specifically to deal with problems after they had already happened—to take preventative measures to stop behavior before it started. On top of that, the agencies tasked to solve such conundrums seemed to have nothing new to say, each piggybacking off the research of its peers rather than developing new interpretations and strategies. This was the fruit of the transitional debates at the Denver glue summit. It was an expanded rumination on Lohman's keynote address, and it demonstrated that new ways of thinking about such problems were steadily beginning to take hold among researchers of model airplane glue. To combat the problem, Kupperstein and Susman published an extensive bibliography of the extant literature on glue sniffing in the United States, which provided a blueprint for administrators, giving them a knowledge base that would allow them to build off of previous scholarship and develop new techniques to combat the epidemic.[32]

But not entirely. When the next glue-sniffing study appeared, its introduction proved just as contradictory and hyperbolic, seemingly ignoring the warnings of its predecessors. Glue sniffers "have used ingenious and novel ways of achieving their aim, [but] it is a fact that a well-balanced, emotionally-healthy individual will not allow himself to even start experimenting with drugs." Contradictions. And sniffing "model airplane cements is probably much more widespread than we dare believe." Hyperbole. But Columbia researchers James Chapel and Daniel Taylor did have some surprising claims to make. They described the immediate and long-term effects of sniffing glue, closing with a statement that hadn't been made in several years. "It should be appreciated that the concept of permanent damage resulting from glue sniffing is opposed by some clinicians," they argued. "However, there seems to be evidence to support the contention that inhalation of toluene is capable of causing permanent organic brain damage and aplastic anemia, as well as liver, kidney and lung damage." What? Hadn't such claims been debunked over the past several years? As it turns out, yes. Chapel and Taylor did not cite their own research in making such claims. Instead, they cited articles from earlier in the decade that had since been called into question by the scientific community.[33]

Such was the problem with the pursuit of the glue-sniffing monster. "Data" was built upon culling the data of others, leaving everyone spinning their wheels. Chapel and Taylor went on to hash out the development of tolerance to model airplane glue. They discussed the possibilities of addiction and the fine points of parsing the difference between addiction and habituation. Their profile of the glue sniffer was familiar, as was their overt worry about the potential of boys under the influence to behave recklessly or violently. They told the story of a fifteen-year-old boy who forced his family out of their home before shooting himself in the leg. A fourteen-year-old shot himself fatally after sniffing. But reckless behavior could also be more subtle. "One of the effects of glue sniffing is described as being a 'dissolver of inhibitions' and, in this context, it would be appropriate to conclude that latent homosexual tendencies might well pop through to consciousness when the inhibitions are loosened."[34] It was presented as new information, with the reputation of Columbia University behind it, but the article itself simply rehashed the kinds of scare headlines that constituted the original research on sniffing glue in the early part of the decade—to say nothing of the sex-laden Freudian theories on the sense of smell from the early part of the century.[35]

But problems still existed on the ground, as well, far from the madding crowd of scientists and law enforcement officers. Pittsburgh's Charles Winek described an early 1968 case where a thirteen-year-old boy died while sniffing glue in his bedroom. He was found twelve hours after his death, his feet singed from their resting place on the radiator. His lungs were congested, there were hemorrhages on his larynx and trachea. His spleen and liver also showed fairly obvious signs of congestion. Ultimately, suffocation from the plastic bag the boy was using to sniff glue did him in, but Winek was more concerned with the organ damage stemming from a toluene-based product. Studies had shown that despite the toxicity of toluene, benzene ultimately proved far more dangerous in the long run. But Winek didn't think so. There were trace amounts of benzene in model glue, existing as impurities in the toluene, which was a methylbenzene. "Toluene and benzene are both considered to be pharmacological narcotics," he noted, "toluene being a more powerful narcotic and more acutely toxic than benzene."[36] It could have been an overreaction to the death. It could have been the cry of "fire" in a crowded theater. Or, the evidence of damage to the boy's internal organs could have convinced Winek that benzene's long-term bodily harm was matched by toluene's short-term bodily harm. And if that was the case, users didn't have to fall off roofs or choke on plastic bags to die from exposure. Again, the problem with such conclusions was that the sample set was so small, any core conclusions could never be mapped onto the entire country.

And besides, "the problem of chronic toxicity resulting from repeated inhalations of the fumes of plastic cements is difficult to assess," according to Virginia's state health commissioner Mack I. Shanholtz. Of course, this didn't stop him from claiming that "glue sniffing is physically harmless to most children." As strange as this statement was, Shanholtz was still worried about model airplane glue and considered it a serious problem. He heartily endorsed a Virginia glue-sniffing statute that became law in early 1968. It banned the inhalation of toluene, benzene, acetone, and other substances in a variety of household cleaning products, glues, and gasoline. Two-hundred-dollar fines and six-month jail sentences awaited offenders.[37] The problem with the Virginia law, which Shanholtz never acknowledged, was that, without regulating and punishing sales of such glue, the legal effects of the legislation would be limited.

This isn't to say that emphasis on the glue sniffer himself wasn't an important function. Shortly after Shanholtz's statements, doctors at Ontario's Ottawa Civic Hospital received an eight-year-old patient with hemolytic uremic syndrome. He often played with model airplane glue and, being constantly surrounded by the fumes in a closed room, often got high from the contact. His blood smear demonstrated leukocytosis. His bone marrow demonstrated signs of acute hemolytic anemia. There was necrosis and petechial hemorrhages on the kidneys and other organs. The boy remained in the hospital for weeks, before dying on his forty-second day of care. Some of the boy's glue sniffing seemed to be intentional, some accidental from contact in his closed room, but the undeniable conclusion was that, despite the statements of Shanholtz, model airplane glue was not physically harmless to children. There was no plastic bag to induce suffocation. There was no homosexual fantasy. The boy didn't walk off a roof. Continuous exposure to model glue—and only continuous exposure to model glue—destroyed his body and his organs' ability to function properly.[38]

Of course, continuous exposure was fundamentally different than occasional, intentional use, but occasional, intentional use would remain the focus of the hunt. Stopping inhalation of model airplane glue had proved problematic for years, and every official and scientist had a different idea of how best to go about convincing adolescents to quit. Perhaps the most effective method was to allow the drugs to do the work for them. John Todd, a doctor at England's High Royds Hospital, described a patient in late 1968 who began sniffing carbon tetrachloride, an industrial liquid to which the adolescent had easy access. His early use was pleasant, but continued use and increasing tolerance ultimately led to "frightening hallucinations." He was surrounded by massive rats, by horses with the heads of dragons—and that kind of disturbing imagery remained constant through several uses. And so he stopped. When the habit stopped being satisfying, there was no longer a reason to continue. Of course, the other consequence of such realities was that if kids could abandon the practice when the high seemed out of control, then it would be difficult to call their dependence addiction, in any of its myriad forms or definitions.[39]

Regardless, Todd's case study was rare, and there were no similar cases of glue sniffers having the same frightening hallucinations. More than likely, when kids stopped sniffing glue, they did so because they had graduated to other substances. And it was "the earliest sign of a child with emotional disorders and anxieties who is turning to drugs for relief."[40] It only got worse from there.

What was really needed was a responsible response to the problem from the HIAA or one of its member companies. And in July 1969, a full decade after the outbreak began, concerned officials finally got it. Testor had been hard at work. Its attempts to discover nontoxic substances that would have the same binding effect as toluene had been unsuccessful, so, as promised, they began in the early 1960s to find additives that might deter use. In 1962 the corporation commissioned an independent laboratory to produce a list of possible chemicals that might make glue sniffing less attractive to American adolescents. The lab submitted a document naming ninety-four chemicals that might possibly do the trick, and Testor began testing them. The grand goal of the company was to find a chemical that would make sniffing the glue undesirable but at the same time wouldn't provide such an offensive odor that model-makers would shy away altogether. Ultimately, they settled on allylisothiocyanate, a "volatile oil of mustard," and in 1968 the company began adding it to its glue products. And it worked. "We can't use Testors anymore," adolescents told a juvenile court official. "They've put something in it and it smelled too bad to sniff."[41]

Allylisothiocyanate was an extract from Indian or black mustard seeds, but it could also be reproduced synthetically. It had "a very pungent irritating odor and an acrid taste." Testor was ecstatic—and eager to make sure that everyone raising hell about the epidemic and its harm to children knew the efforts they had made. "It provides a jolt," Testor stated in a well-publicized news release, "the same as eating horseradish or hot mustard does—and deters people from repeating the action that caused the jolt." It wasn't just a bad smell that left potential sniffers abstinent. It irritated the nose, sinuses, and eyes when purposefully ingested. But model-makers "will not even notice that there was any change in the Testor model cements." Public relations, however, deemed that the corporation should not keep the discovery to itself and announced that it was making the results of its research available to all of its competitors.[42]

Testor milked the development for all it was worth. It was 1969, more than a decade after the first reports of glue sniffing had begun a hysterical reaction across the country. The menace had incubated, grown, matured, and attacked. Now, claimed Testor, the corporation had finally slain the monster.

But for all of its bravado, Testor knew just like everyone else that the monster wasn't dead. The monster, after all, was the fear that model airplane glue created. Science, law enforcement, and state legislators were still hard at work to continue a comprehensive attack on the problem. And no region of the country demonstrated the comprehensive nature of state legislation as a control effort for model airplane glue more than the American South.

# The Law Down South

LAKE PONTCHARTRAIN IS A massive brackish estuary lying along the northern outskirts of greater New Orleans. Second in size only to Utah's Great Salt Lake, it wasn't exactly secluded, but it was far enough away from New Orleans to provide protective cover from the watchful eye of the police—if the group of teenagers needed protective cover at all. They had come out to the lake to get high, but in the warm spring of 1966, they weren't technically doing anything illegal. It wasn't heroin. It wasn't pot. It wasn't even LSD, which would be outlawed by the federal government later that year. The group of Louisiana kids had come to Lake Pontchartrain to sniff glue.[1]

Southern states had been slow to respond to the glue-sniffing problem that had captured the attention and fear of the rest of the nation. The Deep South was a society that had built itself over the last three centuries on personal responsibility and individual rights over and against a legislative program that further impinged on what it deemed to be its personal liberty. The year prior, southern legislators had made much the same case against the federal Voting Rights Act of 1965, vocally decrying federal intervention into a closed system. Glue sniffing and its discontents would provide a trial of the veracity of such claims. Most of the Deep South states would decide to solve the problem with legislation, and regardless of whether legislation was the best method of handling the glue-sniffing phenomenon, the machinations that provided that legislation, as well as the legislation itself, would draw clear distinctions between the states in their thinking about such conundrums and the imperatives that created such thought. In the process, those debates would inadvertently provide an excellent proving ground for examining the minutiae of such legal remedies and the controversies and differences present in such laws.

Of course, despite the variances in interpretation among Deep South legislatures, all of the states in the region debated such a law, and all but Alabama passed one. The process belied the myth of the solid South, but the imperatives that drove legislatures to support such a measure did bridge the significant chasms of race, urban-rural divide, and even party loyalty, and they did so despite the fact that months prior, before the racial defeats of the Civil Rights Act of 1964 and the Voting Rights Act of 1965, such legislation would have been interpreted as the epitome of frivolous government intervention and an assault on the personal liberty and business independence that drove white southern arguments against such civil rights mandates. The Deep South was slow to respond to the glue-sniffing epidemic of the 1960s, but when it did, its state-initiated legislative agenda was far more comprehensive than in any other region of the country.

Even though earlier reports of glue sniffing in southern areas existed, and even though children in the South were far from immune to the practice that had seemingly swept the country, the "closed society" of the protectionist South still in the throes of the civil rights movement and its attendant political upheavals kept such concerns from the front pages of southern newspapers. There was no new social menace more threatening to the area's traditional mores than integration. But as the Civil Rights Act of 1964 and Voting Rights Act of 1965 came and went, proving to the white South that civil rights was an epidemic they would never be able to quarantine, the congressional losses freed southern legislators to focus on the subtler illnesses that plagued the youth of the region. By the time that Deep South legislatures began seriously to emphasize glue sniffing as a legitimate epidemic, myriad municipalities, ten other states, and Puerto Rico had already passed laws regulating the use and purchase of model glue. Significantly, however, conservative western states like Colorado, Arizona, and Utah—genesis points of the epidemic—had not passed laws, preferring instead to let city ordinances in large metroplexes handle the problem.[2] The conservative South, after the defeats of the first-wave civil rights movement had run their course, would be far more proactive at the state level, making a late but concerted effort throughout the second half of the 1960s to stop the epidemic in its tracks. The legal wrangling began on the shores of Louisiana's own great salt lake.

Sniffing glue wasn't technically illegal in Louisiana, but police raided the Lake Pontchartrain party anyway. It was a problem, if not a crime. In April, the state's Social Welfare Planning Council (SWPC) held a seminar in New

Orleans on the developing drug epidemic in the state. Glue sniffing, said the American Social Health Association's Charles Winick, was ominous. "We not only have the immediate problem of the youngster dosing himself, but this may be a prelude to a graver social pathology in addiction to the widely reported instances of death and bizarre behavior, such as walking off roofs and in front of cars." If this weren't enough, Winick told his New Orleans audience, glue could ultimately be a gateway to marijuana, LSD, or heroin.[3] Louisiana's lawmakers were listening.

On May 25, 1966, a contingent of forty-five Louisiana state representatives brought a new bill to the statehouse. It was designed to amend the criminal code to regulate the sale and use of the glue used to make model cars and airplanes. The next day, the proposed legislation went to the Judiciary Committee, and by early June the bill had received a favorable report. It was given a third reading on June 6, then passed (with four votes against) and ordered to the state Senate on June 13.[4]

The bill was spearheaded by New Orleans congressman Eddie Sapir (from the city's Thirteenth Ward), who assured families that children with legitimate interests in models would not be punished. Their parents could buy them the necessary glue. Besides, he argued, both the FBI and juvenile court judges all supported such measures by state governments. Others had already passed such laws. Sniffing glue led to "cruel or violent behavior" and needed to be stopped. "Police records are full of violence and crime because of glue sniffing," Sapir told reporters, "and it only costs fifteen cents a tube. There are no controls now whatsoever."[5]

And controls were precisely what were needed. Three days prior to the passage of the glue-sniffing bill in the House of Representatives, thirteen-year-old Henry Borsch lay dead beside a half-empty can of gasoline in a vacant parking lot in Monroe, Louisiana. It wasn't Borsch's first time, but his worried parents had always managed to find him and stop the behavior before it got out of hand.[6]

While the bill moved through the House, a study sponsored by the city of New Orleans found evidence of increased narcotic use of model airplane glue and other inhaled solvents by children from eleven to thirteen years old, which led councilman Clarence O. Dupuy to sponsor a citywide anti-glue-sniffing ordinance, which passed unanimously. The measure made sniffing glue illegal in New Orleans and placed restrictions on its purchase. It made selling model glue to anyone under eighteen a crime. It was, said Dupuy, "the most comprehensive legislation prepared to date in the United States to combat the ever-increasing menace of glue sniffing to children."[7] The SWPC endorsed the ordinance, as did the Metropolitan Crime Commission and the New Orleans Health Department.[8]

Although the ordinance was passed unanimously, there was still debate. Councilman Walter F. Marcus Jr. wondered about the House bill moving through the legislature and questioned whether the law would override the city ordinance. After all, the city's plan ensured that glue would still be sold with model kits. All sales to minors outside of those bounds would require written consent from a parent or legal guardian. Not to worry, Dupuy assured Marcus. He would ask the Senate to amend the bill to fall in line with Orleans Parish. "This glue-sniffing has become such a menace, that the city just can't stand by and wait to see what the legislature will do."[9]

Sapir, Dupuy, and Marcus were all from New Orleans. Of the forty-five representatives sponsoring the bill, twenty-two were from Orleans and Jefferson parishes. Four more were from East Baton Rouge Parish. Two were from Caddo, home to Shreveport. One was from Lafayette. Eight more were from parishes surrounding those urban areas. Only eight others, in fact, were from rural outlying parishes.[10] The movement, then, was fundamentally urban. Rural areas weren't immune from drug abuse, nor were they immune from kids who spent their free time constructing models. But urban drug abuse was far more prominent, and the pitfalls that accompanied crowded city life made such drug abuse more dangerous, whether that danger came from drug-related violence or from accidents created by intoxication. "Walking off roofs and in front of cars," after all, becomes proportionally more dangerous as the buildings get taller and the traffic becomes more congested.

It would be tempting to argue that rural areas were less likely to support such legislation because of an antipathy to excessive government legislation or because of an antipathy to urban problems in general, but this element of the urban-rural paradigm doesn't seem to be in evidence in Louisiana's glue-sniffing debate. Two of the four House votes against the initial glue-sniffing bill came from rural parishes, but the other two came from Rapides and Calcasieu parishes, home to moderate-sized cities Alexandria and Lake Charles, respectively.[11] Furthermore, when the machinations that created the law ran their course and a final vote was taken, the verdict was unanimous. So the glue-sniffing measure was fundamentally urban because Louisiana legislators saw glue sniffing as an urban problem, but state consent was broad in its application, easily bridging the urban-rural divide.

The principals had something else in common, as well. They were all white. Still, the glue-sniffing measure was supported across racial lines even though this was a South in the throes of civil rights. Founded in 1955 by Ellis F. Hull, the United Voters League (UVL) was one of a series of voting rights organizations in the state, situated among groups like Alexander Tureaud's Orleans Parish Progressive Voters League, the Crescent City Independent Voters League, the New Orleans Voters League, and the New Orleans Voters Association. Its base of support was in the Second Ward, but its influence gave Hull a loud voice. And Hull wanted the glue-sniffing legislation passed. "We the United Voters League, Inc. wholeheartedly support House Bill #752," stated a letter from the UVL to every Louisiana legislator. "Because of a definite increase in the sale of large quantities of glue to minors by community stores and its improper use by buyers, this could lead to the child or children becoming a dope addict; for it is a fact that unrestricted sales of this glue have contributed to juvenile delinquency in the New Orleans area."[12]

Such was the power of the issue. Glue was an intoxicant, and therefore influential with children and teenagers of any race or color. It was cheaper than Schedule One narcotics, so it was available largely to the poor. It was used for model airplanes and cars, so it was also in the hands of the more affluent who could afford such diversions. Glue sniffing didn't have the broader power to unite the races in a time and place of racial conflict. A united white-and-black front against juvenile delinquency did not develop. But the epidemic was able to link the interests of white and black New Orleans parents in a way that was far more visible than it had been previously.

When the state Senate took the bill from the House, its own judiciary committee found it satisfactory in early July, but then floor debate led to a series of amendments proposed by a group of senators led by George D. Tessier, who was (unsurprisingly) also from New Orleans.[13] Councilman Dupuy had instigated much of the amendment talk, but his desire had been to find a "happy medium" between his city ordinance and the House bill. There was, however, nothing medium about the proposed amendments. The House bill defined "model glue" as any substance containing one of twelve different solvents, along with "any other solvent, material, substance, chemical or combination thereof having the property of releasing toxic vapors."[14] It made sniffing or inhaling any such substances a misdemeanor, as it did selling or transferring them to minors unless the donor was a parent or a guardian. Fines ranged between twenty-five and one hundred dollars and up to ninety days in jail.[15]

Senate revisions, however, listed twenty banned solvents.[16] Along with citing the illegality of inhaling those solvents, the bill also made "induc[ing] any other person to do so" a misdemeanor offense. It prohibited possession with intent to use and possession with intent to distribute. Sale to minors had to come with written authorization from parents or guardians, and files had to be kept on all such sales for police inspection for at least one year.

Wholesalers could only sell model glue to retailers "customarily handling such product in the ordinary course of his or its business at a fixed location." And retailers could only sell it if they were recognized as "bona fide" in that custom. Even if retailers did receive their product through reputable wholesalers and even if they had established credentials as the sorts of places that found it necessary to sell model glue, they were still barred from selling more than one tube to any customer, regardless of age, for any twenty-four-hour period. And finally, retailers could not keep model glue on public display, giving easy access to shoplifters.[17]

The state Senate rejected every substantive amendment 24–14 with one absentee. The new version of the bill made model airplane glue sound a lot like a Schedule One narcotic and put a broad range of strictures on small businesses. While paranoia about the destructive narcotic use of chemical solvents was certainly real, the imposition on sellers was more than the body of small government conservative Democrats could take. Still, when the original version of the bill was called to a vote, it passed unanimously. The amendments that did make it through were brief cosmetic and grammatical additions; the House quickly and unanimously approved them and passed the revised bill on July 7. It was signed later that day.[18]

The new law took its place in Part V of the Criminal Code, those "Offenses Affecting the Public Morals." Part V had been modified extensively since the end of World War II. Among the crimes against public morality the revised statutes of 1950 included statutory rape, prostitution, abortion, homosexuality and bestiality (lumped together as "crimes against nature"), and contributing to the delinquency of a minor. Also included was a provision against "cruelty to juveniles," broadly construed.[19] By 1968 prostitution and crimes against adult morality had been excised from Part V and replaced by a new litany of "offenses affecting the health and morals of minors."[20] Selling alcohol and pornography to minors was off-limits. So too was selling them poisonous reptiles. It was illegal to give minors tattoos. It was here, in the set of offenses particularly targeted at juveniles, that the model glue statute resided.[21]

This was an understandable trend, considering that the vast majority of glue sniffers were juveniles. Childhood and its pitfalls were far more dangerous for baby boomers, who grew up in a crucible of student protest, Black Power, and the counterculture that moved around all of it. Still, the placement of the legislation is telling. More and more researchers had been comparing model airplane glue to other, stronger narcotics, but the glue-sniffing ban was not housed with other criminal drug offenses. Heroin, it seemed, could be dangerous to anyone, but sniffing household solvents was something clearly within the purview of adolescents. Here again was the discrepancy between the state Senate's amended bill and the House's original and final product. The Senate amendments made the glue-sniffing legislation into a drug bill. But in the minds of the bulk of state legislators, it wasn't a drug bill. It was a bill promoting "the health and morals of minors." Not all states would view such legislation in the same light.

In practice, however, the law gave authorities guidance as to the arrest and punishment of offenders. The law could work as a drug bill if it had to. Still, the most significant case stemming from the legislation fell wholly in line with the state's thinking. In *Louisiana v. Dimopoullas* (1972), a defendant charged with indecent behavior with juveniles was simultaneously charged with using glue as an accessory to the crime. An adult woman brought model airplane glue to the home of a minor for the express purpose of getting him high, thereby enhancing his willingness "for commission of lewd and lascivious acts." This was transfer without parental consent. It was transfer for the purpose of unlawful inhalation. It was a version of what would later be termed statutory rape. And, ultimately, it was blatant disregard for "the health and morals of minors."[22] Such was the core of Louisiana's thinking about model glue. Had the two conspired with cocaine, for example, the drug charge would have superseded the couple's "lewd and lascivious acts," and more than likely, the teenager would have been seen as a coconspirator. But because the glue-sniffing law aided protective child care over and against the toxin's narcotic classification, the teenager became a victim. *Louisiana v. Dimopoullas* wasn't a drug case. It was a case involving the corruption of a minor. Such subtle differences gave model glue a fundamentally unique standing in Louisiana law. It was a drug, but it was also a tool to corrupt the morality of minors. And although illegal narcotics certainly have the power to corrupt the morality of minors, they don't necessarily have to. And there are plenty of things other than drugs—poisonous reptiles, for example—that can also corrupt their morals. The definition of the law and its placement within the criminal code ultimately changed the criminal meaning of one specific form of narcotic intoxication, making that form of intoxication a fundamentally different crime from all the others.

Of course, the placement of the Louisiana law would also make it fundamentally different from those of other southern states that passed similar statutes. The following summer, for example, on May 9, 1967, Elizabeth J. Johnson, a Republican state senator from Cocoa Beach, Florida, joined with twenty-one of her colleagues—across party lines—to introduce the state's first glue-sniffing measure. Meanwhile, across the hall in the Florida House of Representatives, George Firestone, a Democrat from Miami working in tandem with Johnson, introduced the same measure. HB 1378 and SB 893 differed in name only, in a plan by Johnson and Firestone intended to ensure swift passage of what was seen in a state with a legitimate handful of urban areas as a necessary solution to a growing problem.[23]

Again the movement was far more than urban. Johnson recruited almost half of the state Senate as cosponsors of the bill, legislators representing Miami, Jacksonville, and Tampa, but also Windermere and Altamonte Springs, DeLand and Pinellas Park. There were eleven Democrats and eleven Republicans in the only Deep South state with a significant Republican population.[24] Furthermore, throughout the process of adoption, Firestone would be joined in the House by four fellow representatives, all of whom requested to be recorded as co-introducers of the bill.[25] Again a bipartisan group, again a genuine blend of urban and rural representatives.

Still, the legislature was far from unanimous on how such a law should be framed and administered. When the Senate bill emerged from the Judiciary Committee in early June 1967, it included recommended amendments limiting the ability of retailers to sell more than one tube of model glue to any customer within any designated twenty-four-hour period and barring them from publicly displaying the product on their shelves. Model glue would have to stay behind the counter. The merits of such clauses as preventative measures against possible abuse were obvious, but the budding Sunbelt was less than anxious to place restrictions on businesses simply because it was possible that potential customers might use their products in ways other than those recommended by the manufacturer. Although cosmetic amendments to the original legislation were made and passed later that month with relative ease, the Judiciary amendments went down to defeat. Lakeland's Lawton Chiles proposed striking the suggested additions. The Democrat was not one of the bill's many cosponsors, but he was sympathetic to the cause. Portraying business owners as aiding and abetting, however, would not benefit the growth and development of Lakeland.[26]

The vote was close, twenty-one to seventeen, but the additions were eliminated. Again, the votes did not fall along party lines, nor did they fall along urban-rural divides. There was no unanimity among the bill's sponsors, as they also were divided about whether to legislate the responsibility of sellers. But after the contentious horse trading had run its course, the bill in its final form was read a third time and passed unanimously, forty to zero, with eight members absent from the chamber. Although the role of sellers remained a point of debate everyone, it seemed, wanted to regulate the users.[27]

Meanwhile, the House Judiciary Committee had recommended its version of the glue-sniffing bill a week prior, without any substantive changes.

Later in June, Firestone offered to amend his own bill to match what the Senate Judiciary Committee was seeking in its own legislation. His proposed amendments, along with similar cosmetic changes, included the provisions against same-day sale and public display, and those amendments in the House passed without contest.²⁸

But this was ultimately a Senate project, and Firestone, working in tandem with Johnson, was clearly providing the best backup plan possible pending any setbacks in the other chamber. On July 1, 1967, the adopted Senate version made its way to the House. Although Firestone led the adoption of some further changes to the bill (most notably, a standardized definition of what the legislature meant by the term "model glue"), he bypassed the seller provisions, willingly substituted the Senate version for his own, and submitted the now amended Senate bill for a floor vote in the House. It passed ninety-three to one. The Senate approved the changes, and on July 14 the bill was sent to the governor.²⁹

In its final form, Florida's law listed twenty substances "commonly used in the building of model airplanes, boats and automobiles." It made intentional use of model glue for various forms of intoxication illegal. It made possession or transfer for such purposes illegal. It set a maximum penalty at five hundred dollars or six months in the county jail.³⁰

Congress placed the law in chapter 877 in the Miscellaneous Crimes section of the Criminal Code. Unlike Louisiana's law and those of its fellow Deep South states that would follow, Florida's statute took its place among provisions against tampering with sewer systems, tattooing minors, killing young veal for profit, and mislabeling illegally slaughtered beef.³¹ Florida, in other words, was clear that this was criminal activity, that sniffing glue was both a drug and a phenomenon that was detrimental to the health and morals of minors, but the overwhelming impetus to pass such legislation never translated to a broader discussion about exactly what kind of law the state was creating.

Florida's juvenile courts didn't keep records specifically for glue-sniffing offenses, but they did catalog offenses related to "use or possession of narcotics." In 1965 no such cases existed. This number rose steadily—to nine in 1966 and forty-three in 1967, the year the glue-sniffing measure took effect. The problem, however, is that without any quantifiable measure of which narcotics were being confiscated, the numbers provide little help in demonstrating either a root cause or a justification for the placement of the law. Still, it is significant that "use or possession of narcotics" was in each of those years at the bottom of the juvenile court statistics lists, paling in comparison even to the far more serious crimes of grand larceny and robbery. Such was hardly the stuff of epidemics.³²

Still, the reality of the glue-sniffing menace, as demonstrated in *Louisiana v. Dimopoullas* and countless other frightened newspaper accounts from earlier in the decade, was that it could act as an agent for many of those other, more serious crimes. A 1969 edition of *Florida Health Notes* reminded readers that "a teenage boy who had sniffed three tubes of glue, bashed in the heads of his younger brother, mother and father while they slept." The article posited that the boy was the victim of an unhappy home. Sniffing glue became a method of escape, until it ultimately circled back to create a new, more monstrous set of problems. It was the same kind of applied anecdote that had scared so many others in so many other states. Fitting this kind of analysis, the legislature would respond the following year (1970) with the Drug Abuse Education Act, creating a comprehensive narcotics education program in Florida public schools.[33]

That placement would also have standing in law. In 1973 a juvenile defendant appealed a delinquency conviction on charges of inhalation of harmful chemical substances on grounds that there was insufficient evidence that the chemicals had produced intoxication, but the court ruled that "a chemical analysis was not essential to proof of the charge." In a similar appeal the following year, the court sided with the convicting Dade County Circuit Court, holding that intoxication was sufficiently proven and that "constructive possession" was enough to validate the charge anyway.[34]

As the 1970s progressed, however, with the glue-sniffing "epidemic" functionally over, the court was not so willing to acquiesce. A similar appellant in 1978 argued that his conviction was invalid because the glue-sniffing measure didn't include definitive "warnings of proscribed conduct when measured by common understanding and practice." The vagueness of the law (always a potential problem in Louisiana's legislative debates, as well) made it unconstitutional. It wasn't the first time such a question arose. In 1968 Attorney General Earl Faircloth gave an opinion in response to what exactly the law covered. He argued that model glue "must be a type of glue or cement commonly used in the building of model airplanes, boats and automobiles which contains one or more volatile solvents." The state's worry stemmed from the law's clause adding "any other solvent, material, substance, chemical or combination thereof, having the property of releasing toxic vapors," but the attorney general argued that the addition did not create any new category of toxic substance. It was a catchall, and catchalls never came with the requisite specificity to back formal charges against a defendant. Ten years later, during the 1978 vagueness appeal, the Florida Supreme Court agreed with the appellant. It acknowledged that Florida's law looked similar to other state statutes regulating glue sniffing but also that such similarity didn't warrant sustaining a law that didn't provide specific warnings against improper use.³⁵ And with that, Florida's regulation of glue sniffing came to an end.

Other states' laws would have more staying power. Glue sniffing as a state phenomenon occurred later in Mississippi than it did in Louisiana and Florida. Without the urban hubs of those states, rural Mississippi teenagers were possibly immune from the harsher dictates of the glue-sniffing phenomenon. But more than likely, without the urban dangers that brought such problems to public light, it just took Mississippi authorities a little longer to catch on. The State Department of Public Welfare's youth court statistics didn't acknowledge a glue-sniffing case throughout the early 1960s. Glue sniffing finally emerged as a category in 1967, but it was still far down the list of delinquent acts, well behind theft, disturbing the peace, assault, runaway, sex offense, and others. This didn't make the twenty-three cases that came before Mississippi youth courts that year any less important, but, again, it hardly constituted an epidemic.³⁶

Two of those cases were referred to the court as cases of parental neglect, both by black parents. Of the twenty-one additional cases, thirteen of the juvenile offenders were black, seven white, with one unspecified case. All of them were male. Race, however, cannot be seen as a significant factor in glue-sniffing arrests. In 1968, for example, seventeen of the twenty-two cases were the result of white juvenile offenders. Only five were black. That year did witness the prosecution of two females, but the vast majority of cases were still male.³⁷

Regardless of gender and race, the problem had clearly grown enough to draw the attention of the Mississippi legislature. On Tuesday, February 13, 1968, both the House and Senate introduced glue-sniffing bills and referred them to their respective judiciary committees. The Senate committee was the first to report back and adopted the bill with one revision. The original legislation included a minimum fine of one hundred dollars and a maximum of five hundred to accompany a maximum prison term of ninety days. The Judiciary Committee dropped the minimum fine but otherwise kept the legislation in its original form. When the bill passed the Senate, it moved to the House, whose judiciary committee was still considering its version. The House crafted a compromise measure that kept the elimination of the minimum fine but adopted the measure as House Bill No. 281. Then it went back to the Senate, which pushed the new House bill through Judiciary and passed the measure on July 9. The governor approved the new law on July 12, 1968.³⁸

A number of factors stand out about the Mississippi legislation. First, there were far more votes against the measure than there were in Louisiana and Florida. The Senate bill was introduced by Bill McKinley and Jean Muirhead, both from metropolitan Jackson. They were attorneys, members of various city civic organizations.[39] They had perhaps the best vantage point from which to view the problems glue sniffing could cause. In contradistinction to such advocacy, five senators voted against the Senate bill that moved back to the House before the compromise. Four of the five were from traditionally rural areas. Four were farmers. The fifth, Tommy Munroe, was an oil worker from the coastal military town of Biloxi.[40] These statistics might seem to argue for another clear urban-rural distinction in shaping such policy, but the simultaneous introduction of the House version of Mississippi's glue-sniffing bill makes such conclusions far more complicated.

Of the fifteen representatives who sponsored the legislation, ten of them were from Hinds and Harrison counties, the state's two most populous. The other five legislators hailed from decidedly smaller towns in rural counties.[41] When the bill passed its first time in the House, there were three representatives who voted against it. David Halbrook was from tiny Belzoni and Tommy Campbell was from Yazoo City, but Robert L. Lennon was from the relatively urban college town of Hattiesburg. In addition to such disparities, the core fact remains that the vast majority of both urban and rural representatives, whether lawyers, farmers, or anything else, all voted for the measure in its final form.[42] As in Louisiana and Florida, the impetus for such legislation clearly stemmed from urban areas where the bulk of such problems occurred, but for the most part, outlying areas agreed that the phenomenon required legislation. There are several possible reasons for this acquiescence. First, glue sniffing wasn't solely an urban phenomenon. Second, the bulk of national publicity that the glue-sniffing epidemic received provided an overarching sense that more was to come (or simply scared legislators into submission). Mississippi was late to such legislative imperatives and had been privy to a decade's worth of exposés on the inherent subversiveness of model glue. Third, the youth courts had already been convicting teenage offenders since 1967, so the legislation could easily be interpreted as a validation of current practice or as a deterrent to future abusers. Finally, and as a corollary to youth court conviction statistics, sniffing glue was an instigating agent in myriad other juvenile offenses. Its use made adolescents more likely to disturb the peace, to steal, to fight, to commit various sex offenses. By legislating away one of the causes of such behavior, legislators from all parts of Mississippi hoped to curb the dominant forms of juvenile crime.

With such entrenched imperatives governing the process, votes against the Mississippi glue-sniffing bill seem far more the result of political posturing than of any philosophical problem with the bill itself. When the approved House measure returned to the Senate and the vote was taken on the final compromise, votes in the negative increased to nine. But only Tommy Munroe, the least likely of the original Senate detractors, remained in the core group of opponents. The other four, all rural farmers, approved the final draft of the bill. The eight who replaced them in the "no" column had all originally voted for the bill, and they came from decidedly different backgrounds. Lawyers and farmers from both rural and relatively urban districts opposed the measure. One, H. C. Strider of Charleston, was the former sheriff of Tallahatchie County.[43] All eight had voted for the original bill, and the new version incorporated the one change that the Senate Judiciary Committee made, excising the minimum fine from the legislation. The only difference between the original bills was the designation. The Senate was now voting on House Bill 281, not Senate Bill 1072. While the possibility exists that eight senators from eight different districts each had a significant change of mind about the nature of the glue-sniffing bill, the votes (in what appeared to be a foregone approval) were most likely the result of legislative rivalry.

The bill that rivalry created was remarkably similar to the one created by Louisiana. Mississippi had learned from its neighbor. Not only was there not any substantive debate about the possibility of creating a drug bill along the lines of Louisiana's rejected proposal, there wasn't any substantive debate about the wording of the law at all, with the exception of the penalty clause. In fact, Mississippi's law listed the exact same solvents as did Louisiana's final draft—in the exact same order. Its description of the unlawful act of sniffing glue was the same. Its description of unlawful transfer was the same. With the exception of an omission of Louisiana's caveat that the glue-sniffing statute "shall not apply to the inhalation of any anesthesia for medical or dental purposes" and a slight rewording of the definition of custodial vendors, the 1968 Mississippi law was a direct crib of its counterpart two years prior.[44]

The one substantive difference between the two bills was the value of fines associated with the offense. The Mississippi Senate rid itself of the one-hundred-dollar minimum fine, but Louisiana's maximum fine was only one hundred dollars. Of course, the other principal difference between the two pieces of legislation had nothing to do with the wording of the bills themselves. Louisiana clearly saw its law as an element of child protection, placing it under the auspices of the broader heading, "The Health and Morals of Minors." Mississippi saw its law differently. House Bill No. 281 wasn't a drug bill, but it wasn't necessarily a morality bill either. Instead, legislators placed the bill among "Crimes Affecting Public Health," grouping it with a series of other legislative strictures on poisons—buying or using arsenic illegally, selling poisons to minors, poisoning fish or other animals, poisoning food, drinks, medicines, or poisoning with intent to kill.[45] Glue, in the mind of Mississippi, was a poison, cordoned off through that definition from both juvenile delinquency and drug offenses.

Still, there are no points in lawmaking for originality, and the law's placement in the criminal code still made it valid and (hopefully, thought Mississippi) preventative. Prevention is hard to gauge for a crime that yielded 22 convictions in the year of its passage and 23 the year prior. The small statistical sample makes any real conclusions about prevention impossible to draw—with the possible exception that Mississippi legislators were chasing the shadow of a monster created by the media. But in 1969, the total number of convictions did drop to 15, a statistically significant reduction. In addition, cases of juvenile theft dropped from 2,255 in 1968 to 2,021 in 1969. Sex crimes dropped from 159 to 122. Such reductions cannot be placed solely at the feet of the glue-sniffing bill, but model airplane glue's consistently cited role in the commission of such offenses prior to the legislation indicates that, at the very least, it had an effect.[46]

Mississippi, however, wasn't the only Gulf South state debating the merits of glue-sniffing legislation in 1968. Although Georgia was dealing with the legal ramifications of the menace at the same time as Mississippi, its law would ultimately be very different. As early as June 12, 1966, *Atlanta Journal* columnist Robert Coram wrote an exposé on the glue-sniffing phenomenon in the city.

> What else is there for a 10 or 11-year-old boy to do? Here it is summer time and school is out. They can't drive a car the way big guys do. Going with girls is out. And who wants to hang around the back yard or even a big playground. There's really nothing left to do but take 15 cents, hop on down to the corner store, buy a tube of glue and a paper bag and have a real whingding of a glue sniffing party.[47]

He explained for those uninitiated with the practice how glue sniffing worked, then removed his tongue from his cheek and launched into a blistering assault. There was the twelve-year-old boy who quixotically attacked four marines. Another who jumped from a building window after convincing himself he could fly. Another who lay on active train tracks to prove he was the strongest man in the world.

Fulton County juvenile court judge Elmo Holt saw hundreds of such cases parading through his chambers. One sweep of a local elementary school, in fact, found thirty glue sniffers. Holt's concern about the problem led him to study the behavior of juvenile delinquents in Atlanta. Of the ninety-six boys serving in the juvenile detention center, thirty-six admitted to being glue sniffers, and thirty of those were deemed by staff officials to be "chronic users." It was reasonable to assume, argued Holt, that at least ten more were either afraid or unwilling to admit use. If such numbers held outside of the detention center, "then we're in trouble—real trouble."[48]

On top of its immediate dangers, he argued, glue sniffing surely had more lasting effects. It could cause sterility. It could cause brain damage. It unquestionably did cause skin inflammation. "They come in here with their eyes bugged out, their faces flushed and they don't know what's going on," Holt said. "We've had them in here accused of robbery, burglary, rape and murder. We ask them about the charge and they say, 'I don't know . . . I've been sniffing glue.'" Such psychological effects were even more dangerous. Coram cited a direct link between early onset glue sniffing and later destructive drug and alcohol behaviors, as children—usually between twelve and fourteen—"graduated" to other substances when they became older. Girls weren't immune to such behavior, but it was largely a male phenomenon, and the boys who did sniff glue were generally those with low self-esteem, followers "whose inner nature predisposes them to a less-than-normal interest in the opposite sex."[49]

The article was covering well-trodden ground but at least had the benefit of multiple cases rather than a single anecdote. In place of petitioning for legislation against the practice, Holt sent circulars to stores that sold model glue, warning them not to supply the substance to children without parental consent. The "voluntary support by the merchants" was a key component in controlling the problem, but it was also impossible to regulate such behavior through circulars. Reports surfaced that at some Atlanta stores, children would buy several tubes of glue, and clerks would offer them paper bags to use for huffing it.[50]

And so, decided Holt, the real onus fell to parents. "The underlying problem," he argued, "is lack of supervision at home." Parental neglect or lack of concern only abetted a would-be glue sniffer. Clinton Chafin, Atlanta's superintendent of detectives, agreed. "The problem is getting more serious by the day," he said. "Consequently, this imposes a great responsibility on parents to inform and educate their children on the real dangers involved in glue sniffing." There was, after all, only so much Chafin could do. Glue sniffing was not illegal. Since Georgia's juvenile court had jurisdiction over the "health and welfare of juveniles," judges like Holt could intervene in glue-sniffing cases, but until the practice led adolescents to some other form of juvenile delinquency, the court's hands were tied. And this was the largest problem because, said Holt, in Atlanta, "glue sniffing has reached epidemic proportions."[51]

Still, Atlanta's response to the epidemic was largely to encourage merchant and parental responsibility. Holt acknowledged that if he had known what kind of problem glue sniffing would become in 1966, he would have pushed for legislation in January, but even with the problem growing in front of him, he wasn't pushing for legislation in June. Less than a month later, in response to a Houston ordinance that prohibited the distribution of model glue to anyone under twenty-one, editors of the *Journal* questioned its readers in an editorial. "What kind of a society is it that has to pass laws to prevent misuse of such a mundane item as glue? And we will leave it for everyone present to answer. After all we are that society, aren't we?"[52]

Elmo Holt had done his best in 1966 to instigate a renaissance of merchant and parental responsibility in dealing with the glue-sniffing epidemic plaguing Atlanta, but fellow juvenile court judge Curtis Tillman wasn't satisfied. At his request, DeKalb County passed an ordinance in August 1966 to deal with the "growing menace," allowing for prosecution of persons under the influence of model glue and limiting individual purchases of the substance to one tube per visit.[53] It wasn't a particularly stringent ordinance, but it was a start. The problem was that DeKalb, while part of the metroplex, didn't cover much of Atlanta proper. Any kind of systemic change would need to come either from Fulton County or from Atlanta city officials themselves.

This response would come in February 1967, when Atlanta's board of aldermen passed its own preventative ordinance. It banned the sale of model airplane glue to anyone under eighteen and limited the purchase by anyone else to one tube in any twenty-four-hour period. Sniffing was illegal, but possession was not. "This keeps children who make model airplanes from buying it, but it doesn't preclude possession of glue," explained alderman Richard Freeman. "The child's parents could get it for him."[54] Atlanta's version was similar to DeKalb County's, but it seemed to fly in the face of Holt's original assessment of the problem. In June 1966, Holt argued that a lack of parental concern was at the heart of the epidemic, but Atlanta's February ordinance seemed to place the responsibility for glue use directly in the hands of parents. It was an important effort, but such discrepancies made the bill seem like a temporary fix rather than a final solution.

The same month that Atlanta's new ordinance appeared, the Georgia Department of Public Health featured a cover story in its newsletter on the glue sniffer's "quest for ecstacy [sic]." The intoxicant could cause brain damage or death, but even in cases less severe, glue sniffing created its own form of imprisonment. "With only a brown paper bag and a 10¢ tube of glue a teenager is quickly caught-up in a fantastic world of vivid dreams and hallucinations. Entering this world, like entering Dante's immortal hell, can be a step 'into the eternal darkness, into fire and ice.' It is a world from which some never completely return." It was the kind of hyperbole that had been scaring parents since 1959, but the medical risks were harder to assess. Researchers, after all, had been trying for years. The concentration of various solvents in the glue, the frequency of use, and the physiological differences among users all played roles. Along with brain damage, argued H. K. Sessions, a medical doctor with Georgia's Occupational Health Service, kidney trauma appeared to be the most common medical problem associated with sniffing glue, followed by blood dyscrasis, particularly evinced by bleeding in the lungs. When sniffing glue was combined with other stimulants, the danger only increased. And it was, Sessions argued, like so many of those other stimulants, addictive. He didn't delve into the myriad definitions of what possibly constituted formal addiction, but the numbers were there for everyone to see. Police Superintendent Chafin argued that from January 1966 to January 1967, police arrested 176 glue sniffers.[55]

The Georgia Department of Public Health article in February 1967 closed by describing Atlanta's new ordinance. "Is this the answer? Can legislation prevent teenagers from trying this tempting bout with danger? Will educating teenagers to the permanent damages of glue-sniffing discourage them from trying a fad? The answers are unknown." That they were. But it seemed from available data (or at least from neighboring states) that the best chance at controlling that fad was a combination of both legislation and education. To that end, the *Atlanta Journal* published an in-depth interview with a Vietnam veteran from Atlanta, a twenty-two-year-old who had been a glue addict for the past three years. "Steve," as the article called him, was largely unrepentant about his behavior, though he had been hospitalized and jailed several times. He was temporarily committed to the state mental hospital at Milledgeville. "It's a safer release than a lot of possible releases that society does accept," he argued. But through his defense, the interview sought to use Steve's story as a cautionary tale. He was denied promotion in the military, had lost several jobs. His wife disapproved, and now she was pregnant.[56] The message was clear: sniffing glue could take over your life and destroy it bit by bit.

But the unstated lesson from the article was that glue sniffing was not solely an adolescent and teenage phenomenon—a lesson previously learned at Rikers Island, the University of North Carolina, and the Denver criminal court. Steve was an adult, a veteran. Both DeKalb's and Atlanta's ordinances did include a prohibition on purchasing more than one tube of glue at a time, but other than that, both were directed primarily at children. As was Louisiana's law. As was Florida's. As was Mississippi's. Steve, however, represented a far different demographic, one that largely had been ignored by southern legislators.

Still, the vast majority of abusers over the preceding two years had been juveniles, and there had been a vast majority of abusers. And so, on January 9, 1968, Savannah's Jay Gardner introduced a bill into the Georgia Senate to prohibit distribution and use of model airplane glue for intoxication. As in the cases of other states, the bill moved quickly to Judiciary after being read a second time the following day. But in the intervening time between its introduction and emergence from committee, glue sniffing again entered the news. On Monday, January 15, a sixteen-year-old boy engaged in a firefight with Atlanta police, at one point moaning to draw officers into his home on the suspicion that he had been wounded. They were there to act on a burglary warrant, but when they finally subdued the boy, he had an excuse for the gunfire. "I didn't know what I was doing. [The glue] made me go crazy. I saw visions." Indeed, officers found between twenty-five and thirty tubes of model glue in the boy's apartment.[57] The stakes continued to rise. Glue sniffing, as portrayed in the local newspapers, was becoming more than a truancy instigator, more than a gateway drug, more than a health issue for troubled southern kids. Now it was putting the lives of police officers in danger.

At the end of that week, as the sixteen-year-old boy waited in lockup, the Senate glue-sniffing bill emerged from committee for a floor vote. In its original form, Gardner's bill prohibited sniffing glue, prohibited its sale for illegal purposes, prohibited transfer and sale for those under twenty-one without written consent, and required merchant record-keeping of minor sales. The Judiciary Committee added one substantive amendment to the bill, indemnifying minors who transferred glue to friends for completing models or for other lawful purposes. There would be, in that situation, no need for the minor to keep a record of the transfer or to show written consent before the transfer, as merchants were required to do. In addition, the bill's definition of model glue included far more chemical solvents than any of its southern predecessor bills.[58] The Senate bill was aggressive, emphasized the role of adults, and sought to be comprehensive in its scope. It appeared to be, more than any other bill from any of its neighbor states, a drug bill.

The comprehensive nature of the bill seemed to fit the overwhelming problem stemming at the very least from Georgia's metropolitan areas. When speaking about the bill to reporters, Gardner specifically mentioned adult admissions to the state mental hospital resulting from addiction to solvents found in model glue. The state Senate passed the bill on January 19, 1968, and transferred it to the House ten days later.[59]

The House would seek to rein in the overarching scope of the Senate version. Significantly, the body wouldn't remove its fundamental nature as a drug bill, however. Its Committee on Special Judiciary adopted amendments to the Senate bill that reduced the age of minors from those under twenty-one to those under eighteen. It added a separate provision that assured municipalities and counties that any existing or future ordinances they passed related to glue sniffing would not be repealed by the state bill, providing that any future local ordinances remained consistent with the legislation. The committee also included a new section stating that, in the event some part of the glue-sniffing bill would be declared unconstitutional, the remainder of the bill would remain valid, as "the General Assembly hereby declares that it would have passed the remaining parts of this Act if it had known that such part or parts hereof would be declared or adjudged invalid or unconstitutional."[60] Protecting local government against larger, superseding laws had always been a hallmark of southern legislative thinking, and salvaging parts of legislation against the supposed tyranny of appellate courts had been a lesson hard-learned by southern lawmakers in the preceding civil rights decades. But unlike its Louisiana counterpart, the Georgia legislature did not take the opportunity to negate the broader scope and thrust of the bill, modifying it into a law to govern juvenile delinquency.

And this version passed. Overwhelmingly. The House approved the measure 180-2, but there really is no way to draw any substantive conclusions about the two. Richard L. Starnes from Rome, for example, was the author of Georgia's 1967 abortion statute, which was designed to protect women at risk of death and those who had been victims of rape along with protecting doctors against prosecution for the service. Did his relative liberalism lead him to see this as an unnecessary crackdown on liberty? Did his distance from Atlanta blind him to the urban epidemic? These are questions that cannot be answered. Luckily, they don't need to be answered. The massive overwhelming vote in favor of the measure speaks to the virtual unanimity of Georgia's legislators on the need for the legislation. As in Louisiana, it crossed both gender and racial lines, as the Georgia House of Representatives had been forced to seat eight black legislators following a 1966 Supreme Court ruling. The black congressmen voted for the bill, and everyone agreed—all but two—that the legislation was necessary. When the amended version moved back to the state Senate, for example, it received another unanimous vote. It was sent to the governor in March and signed on April 9, 1968.[61]

Regardless of the near universal support for anti-glue-sniffing legislation, the situation on the ground—in Georgia and across the Deep South—demonstrated that all glue problems were not made equal. Atlanta witnessed a far more dramatic glue-sniffing problem than did rural outlying areas, and so the chain of legislative action began at the local level and advanced to state legislation, hence the provision added by the House to respect existing local ordinances and to allow new ones that didn't directly contradict the state law. This, however, wouldn't last. In April 1970, state attorney general Arthur K. Bolton issued an unofficial opinion declaring that "a city may not adopt an ordinance prohibiting glue sniffing, already denounced by a statute." According to Georgia's constitution, special laws concerning issues already covered by general laws were expressly prohibited. He cited two drunk driving cases from the early 1950s in which the state supreme court overturned convictions under municipal ordinances because the action was already prohibited by the criminal code.[62] Intentions be damned (to say nothing of protections against the encroachment of larger governments against local versions or an inherent fear of the power of appellate courts), that portion of the glue-sniffing legislation would be excised in the early 1970s.

The excision removed one of the principal differences between Georgia's law and those of its neighbors. The remaining differences were still significant. The Georgia statute, in fact, appeared far more similar to the Louisiana Senate's amended bill than to anything that had passed through a southern legislature. It required written records of sales to everyone under eighteen, it required written consent of parents before minors could purchase glue, it prohibited both possession with intent to use and possession with intent to distribute. Like its counterpart laws, the Georgia measure set such offenses as misdemeanors but imposed no minimum or maximum penalties. Finally, the bill did not limit its list of banned solvents at twelve. It didn't stop at twenty, either. The Georgia law listed thirty-four chemicals as ingredients that would make "any glue, cement, solvent or chemical substitute" functionally illegal.[63] Unlike Louisiana, Florida, and Mississippi, Georgia created a drug law.

Consequently, it took residence in the "Controlled Substances" section of the Georgia Criminal Code with other drugs and intoxicants.[64] With the kinds of stipulations enforced in the legislation, and with the kinds of publicized offenses leading to its passage, it was simply impossible to construe the law as one preventing juvenile delinquency. The "health and morals of minors" were important, but Georgia's law was describing something fundamentally broader than that. Poisons, too, required state regulation. But Georgia's law was far more severe than that of Mississippi. When it was being used as an intoxicant, sending users into "Dante's immortal hell," model glue was nothing more than a narcotic that required regulation, punishment, and prevention.

No matter how the House tried to respect local municipalities, this was still a heavy-handed assault on the new drug "epidemic" sweeping the South, and other states in similar positions would cringe at its passage—not because they weren't suffering a similar epidemic, but because of the heavy hands trying to cure it.

On May 6, 1969, well after its fellow Deep South states had passed legislation regulating the sale and use of model glue, Alabama's *Birmingham News* ran an Associated Press story recounting the retirement of Los Angeles police chief Thomas Reddin. "I've about reached the point in my thinking," said Reddin, "where I'm almost willing to write off a generation of young Americans." Reddin stressed communication between police and youth, perhaps a series of programs designed to bridge the chasm of mistrust that had steadily grown between them. Later that morning, Alabama, the most reluctant of the Deep South states to discuss the possibility of regulating inhalation acts, witnessed its first glue-sniffing bill hit the House floor before moving to the body's Health Committee.[65]

It was clear, however, that glue wasn't a priority. The bill didn't receive a second reading for more than a month. Although the Health Committee approved the measure on June 10 and the Rules Committee included it on the docket's priority list the following week, it was allowed to die on the floor without a vote.[66] Such was the fate of many such Alabama bills, and the late consideration and the lack of serious debate might lead one to conclude, particularly in light of the measures in neighboring states that came before it, that Alabama's paranoia about bureaucracy and its devotion to the primacy of local government ultimately won the day. Again, however, such assumptions about a unified South shunning only the bare necessities of legislative need don't hold.

In clear contradistinction to each of its predecessor laws, all of which were sponsored primarily by urban congressmen, the Alabama bill was sponsored by six representatives, with not one from the Birmingham metroplex. John William Grayson was from relatively large Mobile, John Culver and Bert Bank were from the college town of Tuscaloosa, and representatives from Anniston, Gadsden, and Jackson rounded out the group.[67] Alabama's effort was decidedly bereft of such urban imperatives that drove the initiation of similar legislation in similarly minded states. This seeming anomaly, however, doesn't muddy the established paradigm. The precursor Deep South states all came to an urban-rural consensus before passing legislation, and with such a consensus firmly in place authorship of the Alabama law seems almost a fait accompli.

Such is not to say that urban-rural distinctions didn't play a role in Deep South governance. The same day the Alabama glue-sniffing bill came out of committee, a new ad valorem tax bill designed to redefine intangible taxable personal property to include mortgages, domestic stock holdings, and insurance bonds appeared on the floor of the Senate. Introduced by Alton Turner of rural Crenshaw County, it was clearly designed to profit off such urban "property" holdings and thereby equalize the urban-rural share of state property taxes. "If the urban boys can expect us to equalize taxes by assessing a $1,000 piece of land in Crenshaw County at 30 per cent of its fair market value," argued Turner, "then I don't see how they can object to us assessing $1,000 worth of shares in American Tel and Tel [AT&T] at the same rate. That's real equalization, isn't it?"[68] Such rivalries were real, they were significant, but they largely disappeared when it came to race, crime, and child welfare. Their appearance almost always attended discussions of issues specific to the monetary burden of one side or the other.

Regardless, the wording of the state's glue-sniffing legislation mirrored that of its Mississippi neighbor. It outlawed intentional inhalation "for the purpose of causing a condition of, or inducing symptoms of intoxication, elation, euphoria, dizziness, excitement, irrational behavior, exhilaration, paralysis, stupefaction or dulling of the senses of the nervous system, or for the purpose of, in any manner, changing, distorting or disturbing the audio, visual or mental processes." It, too, exempted anesthesia. It banned illegal possession, transfer, and possession with intent to transfer. There was a maximum fine of five hundred dollars or six months in county prison. Its only real substantive difference from the Mississippi law was its list of volatile solvents, which numbered to twenty.[69]

The Alabama bill, however, never saw the light of day. It was the summer of 1969. The hysteria about the "glue sniffing epidemic" was beginning to pass. Testor had already loaded its model airplane glue with allylisothiocyanate. Besides, the Alabama House of Representatives was controlled by the iron fist of Speaker Rankin Fite, a north Alabama politician, close ally of George Wallace, and a power broker who used his influence to pocket bills he thought to be frivolous or unnecessary. In 1967 he created the House Committee on Highway Safety, as the *Birmingham News* explained, "as a mostly inactive depository for bills he hoped to kill."[70]

The House was incensed. As the glue bill came out of committee, freshman representative Charles Wright of Etowah rose to lambast the Speaker for obstructing the business of government. "In this session," he said, "you have not led this House. You have been this House." And no one, he claimed, seemed to be willing to do anything about such tactics because of nothing more than base fear. "We have been in session now eight days and have accomplished absolutely nothing." Fite was able to withstand such challenges, but clearly the idea that the vast majority of legislative imperatives were Trojan horses at best and rank government encroachment at worst was not a uniform opinion.[71]

While the Alabama bill met its end at the hands of a dictatorial legislative leader, the reasons for its failure have to be seen as more than the sum of one man's will. It was 1969, a full decade since the glue-sniffing epidemic had appeared in Tucson and Denver. Alabama's youth were still surely susceptible to such enticements and dangers, but the national headlines warning of dangerous rooftop sex orgies were gone. The disease remained, but the hysteria had largely dissipated. The South's eventual consideration, with whatever deliberate speed it took to arrive at this point, demonstrated that it was perfectly willing to impinge on personal liberty and Sunbelt business imperatives in aid of remedying the ills of its population, functionally nullifying all the arguments it made on the barricades of the civil rights movement. At the same time, however, the state legislators of the Deep South demonstrated in microcosm the kinds of laws that states around the country were producing and the kinds of internal debates that governed how each would ultimately appear.

Those laws would also set the stage for punitive anti-narcotics laws that would define the coming War on Drugs. In fact, in 1965, as Deep South states began paying attention to model airplane glue, the federal government created the Bureau of Drug Abuse Control under the auspices of the Food and Drug Administration, and three years later the organization merged with the Treasury Department's Bureau of Narcotics to create the Bureau of Narcotics and Dangerous Drugs, housed in the Justice Department.[72] Those same Deep South states would support Richard Nixon three years later, responding to his "Southern Strategy" and his racially coded "law and order" language (see chapter 10). States that sought to police adolescent and preadolescent glue sniffing with legislation were more than willing to enforce the legal manifestations of Nixon's harsh assault on drugs designed for older children. There was a massive jump in juvenile drug arrests from 1965 to 1970. Defendants paraded through district courts grew exponentially. As the paranoia over sniffing glue created a precedent for the paranoia over drugs without potentially positive uses generated by the Nixon and Reagan administrations, so too did those state laws trying to police model airplane glue provide the beginnings of a state infrastructure to respond to the more comprehensive federal drug legislation to come.[73]

This isn't to say that these laws were necessarily good ideas. "Panic-driven public spending generates over the long term a pathology akin to one found in drug addicts," argues sociologist Barry Glassner. While state governments spend time and money legislating against compulsions, the time and resources devoted to more pressing matters commensurately diminish. "While fortunes are being spent to protect children from dangers that few ever encounter, approximately 11 million children lack health insurance, 12 million are malnourished, and rates of illiteracy are increasing." As Philip Jenkins has demonstrated, a corresponding moral panic relating to children—that relating to sexual abuse—began to retrench from an overwhelming legalistic approach in the late 1960s, as more and more critics began to question policies that "reflected outmoded sexual prejudices" and imposed severe penalties and intrusive procedures such as electroshock treatment on convicted offenders. The difference, of course, is that the child sexual abuse panic developed in the Gilded Age and went through several phases over the course of the twentieth century. Model airplane glue, however, didn't have that kind of history. It was, to so many who read the public declamations, a new problem with immediate dangers that had to be curtailed.[74]

That reactionary attitude was able to convince the country's most reactionary region. Through the variations in legal intent and scope, each Deep South state—save Alabama—passed legislation that regulated the purchase, transfer, and inhalation of toxic solvents with the aim of protecting the youth of the South. In the process, those states developed the most comprehensive regional push to quell the epidemic, willingly bending their own legislative assumptions to ensure effective control. Such, it seemed, was the only way to protect southern children from "Dante's immortal hell."

# The Lingering Evil

"CAN YOU BELIEVE IT?" *Washington Post* columnist John Carmody seemed to be asking in January 1967. They held a glue-sniffing conference out in Denver and "deplored the lack of communications between adults and the young on the problem." They "referred to glue-sniffing as 'substance abuse.'"[1] These guys couldn't be serious. All anyone ever heard about as of January 1967 was the danger of kids sniffing model airplane glue. And "substance abuse"? Really? What was often lost in the efforts by scientists, doctors, social workers, law enforcement officers, and state legislators to get their hands, minds, and laws around the problem of sniffing glue was the public reaction to their work—and the public reaction to the epidemic on the ground. From the end of the glue summit to the end of the decade, the situation on the ground remained problematic enough to keep the press and public up in arms. They were far from the ivory tower, and the situation as they experienced it still seemed without anything close to satisfactory resolution.

New York, for example, admitted soon after the glue summit came to a conclusion that a full one-fifth of the city's students "are two years retarded in reading." It was a school district in February 1967 with more than a million students and a budget approaching a billion dollars. Maybe it was poor teachers. Maybe it was poverty. Maybe it was immigrant adjustment. And maybe at least part of it was model airplane glue. Principal Ruth Ryan, who had battled sniffing glue in New York public schools, couldn't buy that. But she also couldn't buy that poor teachers were hurting students. "We like to work with the parents," she said. "But some mothers are afraid to come to school, because their homes have been robbed while they were attending a parent meeting."[2] Poverty, went her assumption, was the real problem, and as study after study had shown, glue was no longer a problem of poverty. A Long Island study conducted at the same time demonstrated that at least 6 percent of one particular student body (155 students out of 2,587) admitted to using model airplane glue. Surely not everyone was willing to admit use (approximately 8 percent admitted to marijuana use, almost 2 percent to LSD or some other hallucinogenic), but the survey was anonymous and there was little reason to lie.[3]

Later that month, just down Interstate 95 in Fairfax, Virginia, the local chief of police charged fifteen members of a "glue-sniffing gang" with delinquency. Two of the gang members had become addicted—or believed they had become addicted, depending on the variable academic definitions of addiction—and turned to the police for help after their parents had scolded them without providing any help or treatment. The admissions led to a sting conducted by police chief Murray Kutner, who caught and charged the boys. "What we are worried about is the kids hurting themselves," he told reporters, though the difference between charging the boys as delinquents and scolding them without help or treatment seemed a pretty fine line. Regardless, the kids were charged. Ted L. Rothstein, a medical consultant for the Department of Health, Education, and Welfare's National Center for Chronic Disease Control felt compelled to respond, to provide the information that the boys were apparently not receiving in Fairfax. "There is evidence that the euphoria resulting from glue-sniffing has led to impulsive and destructive behavior," he wrote. "Amnesia has occurred for events that took place during the acute intoxication stage." He discussed the possibility of kidney problems, lung and bone marrow problems. Virginia "has enacted no legislation proscribing the use of toxic inhalants," he chided (Virginia's law would soon appear, in February 1967). Maryland had a law. Washington, DC, had a law. It was necessary. But it wasn't the only thing that was necessary. As the problems in Long Island had demonstrated, the children of Virginia needed to be educated about the dangers of glue sniffing. He knew that people were concerned that more education just meant more ideas for students, but "whatever risks there are inherent in disseminating information about glue sniffing to the public are worth undertaking if the educational material is made effective and can overcome the mass of misinformation that clouds the dangers of this practice."[4]

The educational material continued to grow. In April 1967, the drug company Smith, Kline and French Laboratories, cooperating with the National Education Association, published a small paperback titled *Drug Abuse: Escape to Nowhere*. It was prepared by an advisory panel of scientists, but the book was designed to be a lay tool for teachers and parents. It encouraged early education about a variety of drugs including model airplane glue but discouraged the kinds of scare tactics that had dominated previous educational models. Education about glue and other drugs, it argued, "should be informative and factual. Sermonizing MUST be avoided."[5] To this end, *Drug Abuse: Escape to Nowhere* included a glossary of slang drug terms so parents could pick up on the lingo of their children. Of course, hearing a kid call the police "fuzz" didn't necessarily mean he was on drugs, but it was a useful place to start.[6]

And that was, after all, the principal admonition of researchers, stemming from the glue summit and continuing throughout the rest of the decade. Education was important, even with the inherent risk that children would use the information in a negative way. Along with the book, pamphlets began to appear with much the same rudimentary information, designed to provide parents and teachers with the tools they needed to detect glue and other drug use. Everything was going according to plan. "There's no need for parents and teachers just to stand by feeling helpless or shocked by the growing number of youngsters, rich and poor, who are fooling around with dangerous drugs," wrote *Washington Post* columnist Dorothy Rich. Adults needed to "communicate with their youngsters, to find out and care about how they feel," rather than condemn them in hindsight. To get that done, "experts advise discussions about drugs in advance at home and in school, even before youngsters are tempted to try them."[7] Such had been the call of law enforcement officials and researchers. It was much more convenient to provide education and warnings before children started experimenting with substances like glue than it was to prosecute them after the fact.

The other call from officials was that the glue-sniffing epidemic continued to spread. It did so in a variety of ways, the most obvious being the continued use by more and more children willing to give it a try. As the epidemic had developed through the decade, however, there were other forms of dissemination that weren't always so obvious. First, adolescents often moved from glue to other products in their household that could also get them high. Second, a problem that seemed distinctively American moved to other countries that had formerly assumed themselves to be epidemic-free. And finally, problematic behaviors from children who used model airplane glue became more frightening. As it happened, even in the midst of the new popular educational push, all three of these developments began appearing. Such was the nature of an epidemic driven by fear.

In the spring of 1967, law enforcement officers and fruit companies went public with a new household menace: bananas. Those looking for a legal high scraped the pulp from banana peels, dried it, and then smoked it. A spokesman for the Federal Bureau of Drug Abuse Control admitted that "there is nothing we can do." Unlike model airplane glue, there was no feasible way for the government to make selling or eating bananas illegal. Officials could publicize the problem, but there really wasn't anything more authorities could use to curb it. "And besides," asked the *Wall Street Journal*, "what legislator would dare to affix his name to 'The Banana Control Act of 1968'?" Meanwhile, in June, Japanese police found a fishing boat filled with five dead teenagers, all killed by sniffing a lacquer thinner loaded with ethyl alcohol and toluene. A week later, in Brooklyn, twenty-year-old Domingo Soto was caught sniffing glue with his twelve-year-old brother and his classmate. Police arrested Soto, but he broke away, grabbed a wooden banister, and struck one of the officers in the shoulder. He ran up a set of stairs and hid in a vacant room. When the police found him, he slashed one of them with a knife, leaving the officers with no choice but to fire their weapons and kill him.[8] The transition to other products, the spread to other countries, frightening behavior by those sniffing glue: all continued, standing as testaments to the fact that the confused and often contradictory scientific research about model airplane glue that trickled down to the public and popular press, which sought with best intentions to end the epidemic, in fact only drove it onward.

Interestingly, however, as the glue-sniffing phenomenon continued to escalate, more public worry was devoted to other drugs, particularly marijuana and LSD. Model airplane glue was almost universally listed as a precursor to such substances, but as the country found itself in the throes of the counterculture, the popular hysteria had begun to shift. That counterculture had built itself throughout the decade on opposition to the Vietnam War and devotion to the civil rights movement. From the creation of Students for a Democratic Society at Columbia University in 1962 to the student revolts at Cal-Berkeley in 1965, from Abbie Hoffman's protests at the 1968 Democratic National Convention to the National Guard shootings at Kent State University in 1970, the youth of the country had become politically charged, blaming the bureaucracy of received culture for injustices at home and abroad. Theodore Roszak has argued that much of the countercultural revolt resulted from what he calls the "technocracy," the bureaucratic zeal for modernization and efficiency. "The American young have been somewhat quicker to sense that in the struggle against *this* enemy, the conventional tactics of political resistance have only a marginal place, largely limited to meeting immediate life-and-death crises," he explained in 1969. "Beyond such front-line issues, however, there lies the greater task of altering the total cultural context within which our daily politics takes place."[9] Seeing received culture as the principal culprit of injustice, however, ultimately led many in the so-called student movement to turn against every form of that received culture, including middle-class propriety, monogamy, popular music, and an aversion to recreational drug use.

Thus the American public was exposed to sex, drugs, and rock and roll, none of which in and of themselves had anything to do with a specific political position. Thus countercultural leaders like Ken Kesey and Timothy Leary emerged, emphasizing the benefits of recreational drug use for its own sake. Leary went so far as to create his own church, the League for Spiritual Discovery, in September 1966, for the express purpose of challenging a proposed federal law making LSD illegal on the grounds that the hallucinogen was a religious sacrament. That didn't work. The federal government banned LSD the next month. Pot, barbiturates, and amphetamines were already illegal. But hallucinogenic drugs like LSD, mescaline, and psilocybin were, along with marijuana, nonaddictive substances that led, adherents claimed, to a higher state of being. Thus they became part and parcel of the counterculture, as collegians and twenty-somethings turned on, tuned in, and dropped out, leading unorthodox lives with free love, rock music, and recreational drug use as the cornerstones of the revolt against everything for which their parents' generation stood.

Back in the hot summer of 1967, this meant a string of exposés on the country's drug problem and how it was pulling kids off the straight and narrow path in favor of a life lived without any meaningful cultural production. "If you discover that your youngster possesses or has used drugs, no matter whether or not you may feel personally that marijuana, for instance, is just a 'mild hallucinogen,' the fact is that possession of marijuana and LSD is absolutely illegal and should not be condoned under any circumstances," went one such story.[10] The drugs, the music, the hair. Anti-Vietnam. Anti–college administrator. Anti–white South. The ultimate effect of such associations was that any story about Vietnam protests or student unrest, any story about unorthodox dress or living arrangements, put pot and hallucinogenics into the minds of everyone reading or watching. And so pot and LSD became the new substances that would surely make the sky begin to fall.

Which was to be expected. The problem with such hysteria, however, even when it included model airplane glue as part of the scourge, was that it rarely made distinctions between different kinds of "kids." The average hippie was between eighteen and thirty. The average glue sniffer was between eight and thirteen. There was little question that sniffing glue at an early age could lead kids to other forms of drug use at some point or another, but the two were not mutually exclusive. "Drug use has become so widespread among teenagers from 'nice' families that the experts warn that parents should not only worry about whether or not the teen-ager smokes but also about what he is smoking. It could be marijuana."[11] Such was the kind of rhetoric parents and the press evinced earlier in the decade, when it turned out that it wasn't just impoverished Latinos who were sniffing glue. Advocates for children who used household solvents looked on, hoping that the new hysteria wouldn't completely drown out the old, but it was clear by 1967 that dire warnings about model airplane glue would include dire warnings about the drugs that followed later in childhood. The counterculture, for many crusaders, was the new epidemic.

Still, as summer turned to fall, plenty of glue sniffers remained in juvenile custody (some of them actually escaping from juvenile custody in the Bronx). A Maryland State Health Department report in October 1967 surveyed two "well established, older, white areas" in Baltimore and found that virtually every teenaged boy had tried sniffing glue and roughly 40 percent considered themselves users, sniffing glue at least once a month. And so the Baltimore glue sniffer looked different from the profile created in other cities, "a white, 13- to 20-year-old male, living in an urban area, with poor social adaptation to his family community." The report noted that the older sniffers, sixteen and up, "talk of finding jobs, complain about not finding jobs, and yet sniff glue rather than seek employment."[12] They were tuned in, they had turned on, and they were dropping out. It was evidence of a serious problem in affluent Baltimore neighborhoods, but the hand-wringing and self-flagellation wasn't what it would have been just a couple of years earlier. At the same time that wealthy Baltimore neighborhoods were experiencing a legitimate glue-sniffing epidemic, for example, the affluent Connecticut town of Westport was in the throes of a Methedrine distribution scandal, with several arrests taking place in the five-million-dollar Staples High School. "This is something new," said police lieutenant George Marks. "We've had some glue sniffing and marijuana cases but never anything like this."[13] Methedrine was a version of methamphetamine and certainly a dangerous drug. But what was worse: a handful of amphetamine users in an affluent high school in Connecticut, or the entire teenage population of two affluent neighborhoods in Baltimore sniffing glue? It was a Hobson's choice, to be sure, but the frightening imagery associated with amphetamines seemed to mask the reality that far more students were susceptible to glue. It also demonstrated that the dire rhetoric was shifting to other behaviors. Marks portrayed glue sniffing as a minor irritant that was dwarfed by the frightening reality of harder drugs.

One of the things such comparisons did accomplish, however, was to demonstrate the reality that the use of drugs, and sniffing glue in particular, was no longer a problem relegated to minority boys from broken homes in poor neighborhoods. Interfaith Neighbors, a religious group working to prevent juvenile delinquency, noted that in drug rehabilitation centers that they worked with early in the decade, "there was not a single individual who did not come from a broken home, a home without a father figure or a home in which one or both of the parents were alcoholics. Education, social levels and financial strata made no difference."[14] Obviously, the group's sample size was small and it clearly had an agenda, but it was also no longer working in the early part of the decade. Westport and Baltimore were nothing if not signposts that things had changed.

But they had also stayed the same. In early 1968, a fifteen-year-old Chicago boy died after sniffing glue. A letter writer to the *New York Times* called for contraptions like model airplanes or Freon-based products to be banned as "frivolous yet death-dealing gadgets."[15] Virginia was suitably worried to pass its own legislation (see chapter 7), and Maryland followed close on its heels. And sure enough, such laws seemed necessary. Virginia's statute was scheduled to go into effect in late June 1968, too late to stop a nineteen-year-old drugstore deliveryman in Portsmouth, Virginia, from locking himself in a family bathroom to sniff glue. His parents, after breaking down the door, found him dead.[16]

And then there were the angels. The Young Angels began as a Bronx street gang, but they changed their name in the late 1960s to The Glue Angels, a band of teenagers who sniffed model airplane glue and broke into buildings. Residents were terrorized, and in the summer of 1968 a group of residents formed the Citizens United Organization Association to protect local families and property and to lobby the mayor for a greater police presence in the area. "It is so bad," said a spokesman for the group, "that no one can go out of his apartment. If your apartment is left for five minutes they come in and what they can't take they destroy."[17] The Glue Angels broke the mold of the typical glue-sniffing profile—disorganized loners who dropped out and disengaged. The Angels were a highly organized group of glue sniffers. Sure, they were engaged in problematic activities, but they were engaged. And they weren't loners. Such was the evolution of the problem. Now there were organized criminal enterprises centered around and fueled by model airplane glue. But unlike other evolutionary cycles, this one didn't leave the old profile in its wake. At the same time that The Glue Angels were coming under public scrutiny, a group of Puerto Ricans in Brooklyn were being arrested for glue sniffing while engaged in ethnic battles with local Italians.

Meanwhile, New York's Health Research Council sponsored a study that showed glue sniffing to be one of the most common avenues for eventual heroin addiction. The median age of arrested glue sniffers in New York was fourteen, and almost 20 percent of them were subsequently arrested for use of a different, harder narcotic, usually within one to four years. Officials in Washington, DC, drew similar connections, even noticing that model airplane glue was used in conjunction with or as a temporary stand-in for marijuana and LSD.[18] For all of the emphasis on the countercultural explosion in the United States, the glue-sniffing epidemic was still incredibly prevalent, still incredibly dangerous to users and their victims. And it still evinced far more deaths than marijuana or LSD ever would.

Throughout the 1960s, there was a falling poverty rate and strong school enrollment (see chapter 4), but such numbers ignore the harsh realities of crumbling inner cities and a ballooning federal deficit as every administration since that of Dwight Eisenhower funneled billions of dollars into Southeast Asia to fight the Vietnam War. By 1968, the budget for Vietnam was 26.5 billion dollars and the country's deficit was the largest since World War II. Two years prior to that, a group of concerned activists in inner city Oakland realized the economic and racial devastation infesting the area and responded by creating the Black Panther Party. Martin Luther King, assassinated that April, had devoted himself to the ravaging effects of such inner city devastation. Meanwhile, the federal government took its Bureau of Drug Abuse Control, created in 1965 under the auspices of the Food and Drug Administration, and merged it in 1968 with the Treasury Department's Bureau of Narcotics to create the Bureau of Narcotics and Dangerous Drugs, housed in the Justice Department.[19]

Toward the end of 1968, glue sniffing once again came to the bar in the most influential American case dealing with the epidemic. *People v. Orozco* looked very similar to the 1978 Florida decision that invalidated that state's glue-sniffing law. *Orozco*, however, was decided ten years earlier, and it was decided in California. Two defendants were arrested, tried, and convicted of sniffing glue under the rules of the Monterey Park Municipal Code. But Alfred Orozco and David Valdez appealed. They argued that the statute was not only vague but also invalid because state law preempted it.[20]

Orozco had actually been driving under the influence of model airplane glue before he stopped and passed out. Police discovered him "slumped in the driver's seat while a strong odor of glue emanated from a crumpled cloth in his lap; he appeared to be intoxicated." Meanwhile, his passenger, Valdez, "was found unconscious on the front seat, his face on a diaper which was saturated with glue." Monterey Park, California, had already passed a glue-sniffing measure, banning the inhalation or drinking of "any compound, liquid, chemical or any substance known as glue, adhesive, cement, mucilage, dope or any material or substance, or combination thereof with the intent of becoming intoxicated."[21] Orozco's lawyers argued on appeal of his conviction that the law was so vague that it could apply to smelling any substance, and therefore it didn't adequately inform residents about what was and was not illegal. But the California appellate court wasn't going to allow Orozco to push through a loophole that didn't actually exist. "When considered in the light of the evil which prompted the enactment of the ordinance and the method of control which the city council chose, it is apparent that the legislative purpose was the control of inhalation of glue, not of intoxication in general."[22] It was an important point, because state law dealt with intoxication in general, and that perceived contradiction gave Orozco the legal standing to appeal. After the original conviction, the state of California added an amendment to its drunk-driving statute to include the influence of "toluene or any other substance defined as a poison," but in lieu of that addition, Monterey Park's legislation held firm.[23] Maybe even more noteworthy than the court's reasoning was the court's language. Sniffing glue was evil. It was the kind of language that was a hallmark of the early decade—the fuel of the epidemic itself—and it virtually guaranteed that any glue-sniffing defendant would find no sympathy in a California court.[24]

Sniffers were still finding no sympathy from researchers. A Canadian research team studied a series of teenage glue sniffers and found significant chromosome damage. The study appeared in early 1969, and the popular press ran with it. Unlike so many of the esoteric debates that seemed to stem from researchers on the state and nature of model glue use, chromosome damage was something that everyone could understand. It ensured that "parents have a new worry about the teenage craze of glue sniffing." And even more worrisome, "most of the subjects were not heavy sniffers. They used solvents or glue to get 'high' once or twice a week." Whether or not "twice a week" was "heavy" use was debatable, but chromosome damage really wasn't. And the worry only grew with every new sniffer. South Dakota Sioux tribal judge Hobart Keith acknowledged that juvenile glue sniffers routinely passed through his courtroom. "These Indians better learn to swim or they're going to drown." Even Ann Landers responded, classing model airplane glue with barbiturates, marijuana, LSD, and heroin.[25]

Ann Landers and South Dakota tribal court judges weren't exactly staples of the countercultural movement, but classing glue as being a constituent part of that movement wasn't a mistake. At the 1969 convention of the National Association of Secondary School Principals, leaders noted that almost 60 percent of the nation's high schools had been witness to some form of protest, ranging from vocal complaints to riots. Among the issues students protested? The right to sniff model airplane glue. It wasn't as common an issue as the right to dress as students pleased, the right to smoke, or the problematic scheduling of sporting events, but it was there. Those students weren't dropping out, of school or politics or anything else. And they were acting as a group.[26]

The stakes continued to rise, even as the public divided its drug concerns into all kinds of new elements. To this end, governments like that of Virginia sought to redraft their glue-sniffing legislation to brace for a new round of problems. In Virginia's case, the desire to respond quickly to the crisis almost led to disaster, as the revision "had to be withdrawn for redrafting when it appeared that it would have inadvertently sent anesthesiologists to jail for giving ether to surgical patients."[27]

Even in the face of such changes, however, the glue sniffer profiled so commonly in the first part of the decade hadn't gone away. He wasn't able to take part in a protest at his high school, though, because he wasn't in high school. "Jim," a sixteen-year-old Florida boy, for example, was instead living at the Florida School for Boys, a Marianna correctional school for wayward teens. After bouncing around from foster home to foster home, and running away from many, Jim was first sent to Marianna at age twelve. Before his release he was sexually abused by his fellow inmates, beaten by guards, and placed in solitary confinement. But he was back again at sixteen for sniffing glue. Then he was sent back to solitary for finding gasoline on the grounds and sniffing that too. And this was the point. Reform schools like Marianna didn't work. Vocational training, actual assistance, and the kind of help that could actually turn such delinquents around were nowhere in sight. The problem of juvenile reform was a problem of definitions. "Many people today, including a number of highly successful politicians," wrote Joel Fort, "discuss American youth as though it were one homogeneous entity of tens of millions simultaneously smoking pot, dropping acid, copulating in the streets, and rioting." Sniffing glue and other delinquent behaviors were generally seen as the disease rather than the symptom of adverse social and family conditions. Solitary confinement, beatings, and other horrors faced by juvenile delinquents while in custody managed to avoid fixing both the disease and the symptom. Any kind of real reform—that would provide the help required by wayward glue sniffers in order to have complete and productive lives—would need to start with its definitions. Purging the concept of solitary confinement or punishing guards who overexerted their authority was a good start. Bettering the condition of juveniles, rather than simply punishing them for delinquent behavior, could begin only when those in charge of teenagers like Jim acted as an outlet and a psychiatric aid, acknowledging that glue sniffing was the result of other personal crises, which, when solved, could eliminate the perceived need for such behaviors.[28]

"Youngsters are quite disillusioned, and I think with good reason," said the National Institute of Mental Health's Robert Peterson. "Take a look around you. They are the product of a violent society, one in which there is the balance of terror. They can look forward to fighting and dying. Is the important thing simply material abundance, huckstering?" The crackdown on drugs, glue or otherwise, was useless without fixing the root problems of such behavior, he argued. But then that could be more difficult than it sounded. "No one starts out dependent," said psychiatrist David Trachtenburg. "Everyone starts out for kicks. It seems to me the danger of any of the kids involved is a very subtle thing. Some kids (who abuse drugs) have very little going for them—psychological problems, chronic depression." That was the profile of the early decade, after all. But the profile had been changing. "Others start out apparently having a lot going for them, and then you gradually see their performance deteriorating. I think it goes both ways. You only find out after the fact which kids didn't make it."[29]

This kind of message persisted, even through the decade's many research advances. That same year (1969), Herman Land published a book titled *What You Can Do about Drugs and Your Child*, complete with a foreword by Dr. Henry Brill, who had been on the front lines of the glue-sniffing epidemic since early in the decade (see chapter 5). "This book is written for parents, by a parent," Land opened. His chapter on model airplane glue warned that this particular substance was different from other drugs because it was under the purview of very young children, often as early as the third grade ("yes," he warned, "we are speaking about children at even that tender age"). The volume rehearsed the same kind of frightening arguments that appeared earlier in the decade and included advice to parents upon discovering a glue-sniffing child such as, "Don't strike him," and "Don't have him committed anywhere." Brill's preface, which should have known better, argued that "only a few years ago it was virtually unknown for teenagers to take drugs for pleasure."[30]

As if the news didn't seem dire enough, the Food and Drug Administration claimed in the summer of 1969 that forty-two adolescents had died from inhaling spray products like mouthwash, cleaning fluid, and hair spray. "It causes some dizziness and some elation with one breath, but if they take two or three breaths it freezes their esophagus, congests their lungs, causes asphyxia. It kills."[31] Glue-sniffing deaths weren't included in the total, the spray products being seen as one of the alternatives to the more common model airplane glue. Besides, as of mid-1969, at least some of the glues just couldn't give them the same effect. "You can be sure it's the model—not the kid—that's gonna fly," a Testor advertisement proudly proclaimed.[32] "Because you can't sniff Testor's, there's something in it." The company did everything it could to publicize its addition of mustard-based allylisothiocyanate to its products (see chapter 7). Even John Anderson, the Illinois congressman who represented Testor's hometown of Rockford, announced on the floor of the House of Representatives that through Testor's work, "the problem of 'glue-sniffing' can be eliminated." The company was forced to admit that the addition did cause an initial drop in sales, "but by the end of the year we think sales will be higher than they ever were."[33] President Charles Miller claimed that sales to glue sniffers were always just a small portion of the total anyway. He cynically noted that other household products were still readily available to get kids high (spray products, for example), but now that his company was out from under the gun, he boldly chided manufacturers of household cleaners and other solvents that it was time for them to take the same initiative as the moral and socially responsible Testor Corporation. It seemed easy to be cynical about Miller's braggadocio, but he did make the company's research available to its rivals. At that point, Testor could afford to be magnanimous. The glue-maker told its retailers that they could now sell its products with "confidence that you're promoting the right way for a kid to get off the ground."[34]

But was it the right way? *Washington Post* columnist and renowned geneticist Joshua Lederberg was concerned. He applauded Testor for making the effort but reminded his readers that, twenty-five years prior, mustard oil had been found to cause "lethal mutations in the sperm cells of fruit flies." His argument wasn't that allylisothiocyanate should be removed from model airplane glue. Stopping that menace was a valiant effort. "But we will be plagued by serious and inescapable doubts," he said, "until we can properly attend to the thousands of products that we encounter in our daily life about whose biological effects we know next to nothing."[35]

Even the solutions were fraught with problems. Less than a month after Lederberg's editorial appeared, on November 6, 1969, the federal government once again amended the Federal Hazardous Substances Act, this time attempting to protect children "from toys and other articles intended for use by children which are hazardous due to the presence of electrical, mechanical, or thermal hazards." But regardless of whether children were at risk from electric shock, sharp edges, or potential burns, the new additions were designed to regulate those products that could cause such injuries in the course of "normal use." Again model airplane glue and other household solvents slipped through the federal legislative cracks. The "normal use" of model glue precluded shocks or burns. It could temporarily and frustratingly bind your fingers together or affix a wayward model piece to some article of clothing. But its only real danger lay in instances that could in no way be considered "normal use."[36]

Meanwhile, cities and municipalities were still passing ordinances, parent and teacher groups were still wringing their hands about model airplane glue, and even American GIs began demonstrating disturbing addiction problems. All such reports placed model glue within the context of the country's swelling "drug" epidemic. The contact trace map had fundamentally changed. This time it wasn't the various dots that forced the evolution, it was the lines that connected them. Glue was part and parcel of a new epidemic that included marijuana, LSD, and heroin. But when such different substances, with such different potential users, were lumped together as a "drug" problem, the prospects for coming up with adequate solutions were reduced to virtually nothing. Research had demonstrated that glue sniffers were somewhat likely to later move on to other drugs like heroin, but this shifted American thinking, wrongly, to assume that the markets for glue and heroin were fundamentally similar. A cautionary tale for this kind of thinking came from Japan, whose own (largely successful) war on marijuana had convinced teenagers seeking a similar high to experiment with model airplane glue. To horrifying ends. In 1969, 161 Japanese teenagers died from problems related to sniffing glue, more lives lost than the entire total throughout the American epidemic.[37]

Concerns about the menace of model glue had transmogrified into concerns about the menace of recreational drugs. Glue still lingered as a real problem, but the hysteria that surrounded it had dissipated, diluted as it was in the broader "drug" calamity. It was this kind of watering-down that kept Alabama from passing a glue-sniffing law (see chapter 8). At the same time, it was that same watering-down that left 161 Japanese teenagers dead. The lingering evil of glue sniffing was still screaming for attention, but as the 1960s came to a close, it began to be drowned in a cacophony of other voices.

# The Children's Crusade

MICE DON'T HAVE A choice as to what they sniff and what they don't, particularly those unlucky enough to live in scientific labs. A group of forty of those mice living at Chicago's University of Illinois Hospital inhaled glue fumes for ten minutes. Half of them developed an irregular heartbeat and "a complete blockage of blood flow from the lower to the upper chambers of the heart." All of them died.[1]

The tragic mouse massacre occurred late in 1970, and with more than a decade of similarly tragic human deaths from glue sniffing, researchers were able to draw some new conclusions. They studied reports of 110 sudden deaths following inhalation of model airplane glue, aerosols, and other solvents that had occurred during the 1960s. "We postulate that some humans who sniff glue or solvent vapors die suddenly from ventricular fibrillation, heart block or acute ventricular failure, alone or in combination."[2] It was a significant find, and one that went a long way to explaining fatalities across the country. The researchers made no claims about long-term effects, organ damage, or cognitive brain function. They stuck to what they could prove. And for those who never lived to worry about long-term damage, there was now some solid data to explain why they might have died.

The data also demonstrated that for all the watering-down of the message in the face of countercultural drug use, glue sniffing was fundamentally unique. It was still a problem all its own. This, too, was the message of Ted Rubin, who had manned the barricades of the epidemic since its inception.

The Denver juvenile court judge was also dismayed about the shifting whims of the country's hysteria. To this end, he published one final report, one last shot across the bow for glue-sniffing knowledge, one more missive from Denver, the epidemic's ground zero. His presentation was a distillation of the same Denver position that had existed for the past decade. There could be serious long-term consequences, but no one could be completely sure. Researchers were still working on the problem. Glue sniffers started as early as seven years old. In Denver, they were almost entirely Latino. In Salt Lake City, they were almost entirely white. Black users made up only 4 percent of the total. There was poverty, broken homes, poor grades in school, disastrously low intelligence quotients. Although sniffing glue was a deviant act caused by individual problems, it was still "generally a peer group activity."[3] He discussed counseling and detention and education and all the methods that could be used to wean children off such dangerous substances. As the *Christian Science Monitor* had previously noted, any of these methods, and detention in particular, had their own virtues and vices. More care was needed in order to treat glue sniffing as the unfortunate outgrowth of problems with much deeper roots.[4]

Rubin's newest and most helpful addition to the usual Denver position was a discussion of the legislation that had developed around glue sniffing in states and municipalities across the country. He divided the various laws into three categories, those that "make it illegal to (1) Smell; (2) Sell; or, (3) Smell and Sell." Those laws that made sniffing illegal were relatively straightforward as to what the crime was, but they varied differently on punishment (see chapter 8). "The most notable example is in New York State," Rubin explained, "where a boy between 16 and 21 who sniffs glue can be jailed for three years, while a girl in the same State can be incarcerated for only a maximum of five days." This kind of thinking jibed with the reality that boys were the principal offenders when it came to sniffing glue, but it seemed dramatically unfair. There had been plenty of examples over the preceding decade of girls participating in the activity, but the punishment imperatives seemed to argue that their offenses were somehow less important, less dangerous, than those committed by boys. The gendered nature of the law was rare. No girl, for example, would receive lighter treatment for murdering a neighbor, even though statistics demonstrated that males committed the majority of American murders. Most importantly, however, "restrictive legislation does not remove the root factors which cause habituated solvent sniffing." And juvenile courts weren't really prepared for dealing with such cases anyway. "We can predict failure when traditional probation methods are applied."[5]

Statutes that dealt specifically with sales, on the other hand, didn't demonstrate the same kind of punishment imbalance but they did show a marked difference as to what constituted an illegal sale. "No sale to anyone under a specified age; no sale unless there is also sold within the same transaction a model airplane or boat; sale only when accompanied by a parent or a note from the parent; or the glue must be labeled, warning of the dangers involved before sale is permitted." As demonstrated in chapter 8, some states also included a clause about possession with intent to distribute, which would allow law enforcement officers to punish sniffers who carried several tubes of glue. The formulations mentioned by Rubin, however, dealt specifically with businesses that either retailed or wholesaled the product. Such statutes were probably far more effective in prevention—making it more difficult for would-be sniffers to obtain model airplane glue—but they also placed the burden for the behavior of children on shop owners who were selling products intended to benefit their lives.[6] Certainly there were plenty of hobby shop salespeople who sold model glue with a full understanding that their customers were going to abuse it, but for the vast majority of those salespeople, such laws put a criminal burden on people who were not criminals. Such legislation was almost definitely necessary to help quell such behavior, but criminalizing activities that were not in and of themselves criminal was always a thorny proposition. In addition (and as law enforcement officers had complained about before), it was almost impossible to enforce, as police forces spread thin anyway were now tasked with checking the sales records of hobby shops on the off chance that they might have sold one of their products to someone under age. It was a fool's errand, and even in the places where such laws existed, cops weren't making the herculean effort it would require to make such laws effective.

"Smell and sell" legislation was just a combination of the two and so carried all the same problems that individual smell or sell laws had. When these statutes were enacted by municipalities, the problems were compounded. Denver, for example, had a strong "smell and sell" statute, but none of its neighboring counties, to say nothing of Colorado itself, had similar legislation. "Denver youngsters are therefore free to travel a few miles and sniff without legal consequence." This being the case, Rubin encouraged any new legislation to take the form of a "sell" statute at the state level.[7]

But if making such legislation comprehensive was the great goal, then federal action also had to be considered. It was the next logical step for the problem that still festered even after the public outcry had largely dissipated. And it would ultimately be the last gasp for those on the front lines of the epidemic.

Patsy Takemoto wasn't an unlikely heroine for the glue-sniffing cause. Born in Paia, Maui, Hawaii, in 1927, she grew up on the island to become valedictorian of Maui High School. She began college at the University of Hawaii, but her ambitions grew beyond the bounds of the Pacific Ocean. She bounced around between schools in Pennsylvania and Nebraska before returning to her home island. Still, her penchant for activism was decidedly shown before her return, as her letter-writing campaign against the University of Nebraska's discriminatory housing practices ended in a campus-wide policy change. But then she was gone, back at the University of Hawaii, where she earned degrees in zoology and chemistry before moving to the University of Chicago for her law degree. While there, she met and married a geophysics graduate student named John Francis Mink, and after the completion of their degrees and the birth of a child, the family moved back to the islands.[8]

There were more fights for her back home. She had to work to convince the bar that she was in fact a legitimate resident of Hawaii after her collegiate travels through the United States. After winning that debate, she found herself unable to find a job at any of the island's established law firms, all of whom doubted the ability of a married mother to handle the workload. So she started her own practice. Her success led her to serve as a Democrat in both the Territory of Hawaii House of Representatives and Senate. After Hawaii earned its statehood in 1959, she served in the state Senate. After reapportionment in 1964, Hawaii received an additional seat in the American Congress, and Patsy Mink took it. She would serve in that capacity until 1977, working diligently for women's and civil rights. She was a key supporter of Title IX of the Higher Education Act, and she authored the Women's Educational Equity Act.[9]

She also had a vested interest in model airplane glue. Not only was she a mother who was sensitive to the needs and protection of children, but she had witnessed firsthand the growing menace of glue sniffing in Hawaii. Unlike other activist parents in Congress who served states with prominent glue-sniffing problems, however, Mink was ready to act. On July 10, 1969, she introduced House Resolution 12751, "a bill to amend the Federal Hazardous Substances Act to authorize the Secretary of Health, Education, and Welfare to ban glue and paint products containing toxic solvents." She strode to the podium that afternoon and looked out searchingly at her colleagues. "The deliberate inhalation of vapors from these products is a serious problem in Hawaii and the Nation," she told them. She acknowledged that some of the information as to long-term effects was inherently vague and sometimes contradictory, but "it is clear that this is a highly dangerous practice." Mink submitted her bill just weeks before Testor unveiled its mustard-based solution to the problem, and she spent much of her time maligning the ineffectiveness of corporations under the banner of the HIAA who had yet to find a suitable substance to make model airplane glue less desirable. Then she closed with data from her home state, which had witnessed a dramatic spate of the epidemic. She quoted liberally from Christopher E. Barthel, a mental health consultant to the state's family court system, who expressed his frustration that the national focus had turned to marijuana and other drugs. It seemed to Barthel to be a zero-sum game. The more people fretted over the abuse of marijuana, the less they cared about model airplane glue. "Maladjustment" could clearly occur after use of either substance, but glue sniffing "constitutes a problem worthy of considerably more attention in Hawaii than does the use of marijuana." Glue sniffers tended to be chronic users, whereas marijuana users tended to be experimenters. Glue sniffers came from a significantly younger age group. And perhaps most important, "four times more sniffers than smokers pass through the court" in Hawaii. Barthel also noted that glue sniffers more often came from impoverished areas of Hawaii, from broken homes filled with far more problems than the common marijuana user had to face. That austerity, that powerlessness, if nothing else, made the problem significant and worthy of the kinds of concern that it received throughout the bulk of the decade. Mink placed all such assessments into the congressional record, then ceded the rest of her time, hoping that the information would remind her colleagues that the drug use that had become the hallmark of the counterculture movement did not erase the adolescent glue-sniffing problem that had plagued the country and her state for much longer.[10]

In February 1970, the Interstate and Foreign Commerce Committee's subcommittee on public health and welfare held hearings on potential drug abuse control amendments. Mink's bill wasn't technically part of the Drug Abuse Control Act, but she and her bill were included in the hearings because of the status of model airplane glue as an abused substance. Her testimony acknowledged that "most of the publicity seems to concern marihuana, heroin, and other substances such as LSD." But "the deliberate inhalation of vapors from various solvents, such as those used in the manufacture of glue, is causing great damage to the health of thousands of children across our Nation, and their story is not being told." And children were the thing. Much of the hullabaloo over marijuana and LSD stemmed from the adults who used those drugs, who challenged the idea that they were inherently destructive and consistently petitioned their legislators for the drugs to be legalized. Model airplane glue didn't receive that kind of treatment because

(a) the users were much younger, and (b) model airplane glue didn't need to be legalized. It was the kind of insidious product that was already legal. This was, in fact, what made it so insidious in the first place.

She cited some of the existing research on the subject, then detailed her quixotic efforts to get information from the glue manufacturers about what they were doing about the problem. They were continuing to do research, they told her. About a month before she originally submitted her bill, the HIAA had told her that "we are no more in a position to create a solution than other research bodies are able to create the answer to the existence of cancer." It was a bizarre and inaccurate statement that belied hobby manufacturers' role in profiting from the epidemic throughout the bulk of the 1960s. And so she submitted HR 12751, which would give the Secretary of Health, Education, and Welfare the power to classify model glue as a banned substance unless the company making the product "either (1) manufactures his product without such solvent, or (2) manufactures it with a substance having an obnoxious odor." The Hazardous Substances Act already gave the secretary the power to remove hazardous substances from the market, it was just a matter of putting model airplane glue on that list.[11]

And now the hobby corporations had found a successful additive! This being the case, "there should be no objection by all responsible manufacturers of paint and other products containing similar solvents to a requirement that all of their products be required to contain this obnoxious additive so that our children can be protected." It seemed like a no-brainer. Representatives of the hobby lobby, she claimed, should have no problem with her bill now that there was a scientific solution to the sniffing problem. What was more, this kind of law would protect the entire country. There were so many parts of the nation without glue-sniffing statutes, and those that did have one had taken a "misguided approach." What really were misdemeanor fines able to do? "Why should our innocent children have to pay the price for the negligence of society in permitting free access to these substances, without any form of protection?"[12] It was the great lesson of civil rights. Federal action—not the action of individual states—led to tangible, permanent solutions.

Mink's calls for federal action, however, fell on deaf ears. In March 1970, she stood again at the podium to place her subcommittee testimony into the congressional record. She applauded Congress's care and concern about drug abuse but reminded them that they were missing one incredibly important aspect of the problem. "I refer to glue sniffing, or the wider field of inhalation of fumes from glue, paint, gasoline, and a wide variety of other products containing toxic solvents." She couldn't seem to make them understand that sniffing glue was a fundamentally different problem from smoking marijuana or dropping acid. The nature of the high was different. The availability of the product was different. The age and socioeconomic status of the user was different. And this meant, she argued, that emphasis on the morality crisis brought about by the surge of the counterculture had no bearing on the status of model airplane glue as a danger to the country.[13]

To support her claim, geneticist Joshua Lederberg wrote to John Jarman, chair of the subcommittee on public health and welfare. Lederberg had won the Nobel Prize in Medicine in 1958 and was by far the most eminent voice to weigh in as to the scope of the glue-sniffing problem. "The unhindered availability of solvents does pose a very serious problem in protecting juveniles from very serious injury to themselves with respect not only to behavioral aberrations," he wrote, "but also well documented odds of serious damage to the liver and other internal organs." Lederberg acknowledged the work of Testor and other corporations to add an irritant to their products but argued that such work was even more of a reason to support Mink's legislation. First, there was a lack of "sufficient reliable information about any compound now proposed for use as a deterrent to be sure that it does not become a serious environment pollutant and end up possibly causing more damage than the situation it is supposed to remedy." But this was only one of the problems. Any kind of irritant in model airplane glue, while helpful in keeping children from sniffing it, would almost certainly have the ancillary effect of driving those children to other household solvents. Organizations such as Testor had other corporations within the scope of the HIAA at their disposal and willingly shared their research, but it was unlikely that home cleaning companies or oil companies were going to have much interest in what Testor was doing. If diffusing the threat from model airplane glue was only going to spread the danger to a variety of other products, then the regulation of such products was all the more necessary.[14]

But Lederberg was a scientist, and the bulk of his concern fell on allylisothiocyanate, the mustard oil added to model airplane glues. "The application of mustard oil to the skin and by inhalation together with any of a variety of organic solvents present issues of possible toxicity which are not fully answered by studies on a dietary component which enters the body under the influence of digestive processes and in combination with a very different set of other materials." He was referring to the fact that mustard oil was a component of some foods, most notably horseradish sauce. That, however, didn't make it healthy to the touch, to the nose, or to the eyes.[15] It was a decidedly rational argument from an eminent scientist, but it fell on the ears of a subcommittee with so many other frightening drug menaces around them that, even with the voice of a Nobel Prize winner in their ears, it seemed ultimately frivolous.

Mink's children's crusade, her last-ditch effort to generate a federal statute for the regulation of model airplane glue, was the exclamation point at the end of the decade's horrifying, hyperbolic glue-sniffing sentence. But the paragraph continued. The epidemic had run its course, largely because the epidemic itself was a cultural construct that the mainstream media had functionally abandoned, but the use of model glue and the dangers that accompanied it still existed, still resonated, even if fewer people were listening.

In June 1970, a few short months after Patsy Mink made her last effort in the House of Representatives, the Court of Special Appeals of Maryland ruled in another significant glue-sniffing case. This time, James Joseph Ward, a sixty-one-year-old adult, rented a room at Baltimore's Duke's Motel. On November 12, 1968, he used it as a home base for inviting underage girls into his room to sniff model airplane glue. He was charged under an obscure Maryland law barring "keeping a disorderly house," which served as a kind of catchall to punish anyone using a residence to commit "acts prohibited by statute" or acts that were a nuisance to those around the residence. Sniffing glue was prohibited by statute in Maryland, so Ward was arrested and convicted of a crime. He appealed on two fronts. First, the disorderly house statute was too vague to be meaningful. Second, and most important to Ward's appeal, he hadn't sniffed any glue. Here again was a shining example of Mink's claim that a broad federal mandate was in order. Maryland made it a misdemeanor "for any person to deliberately smell or inhale" model airplane glue for the purpose of intoxication. But Ward didn't sniff glue. He brought girls into his room so that they could sniff it.[16]

Of course, at the time of his arrest, Ward had claimed that he kept the glue for making models in his spare time, and that the girls had come simply to watch television. The arresting officer testified that the room's glue fumes "hit you in the face" upon entry. It was damning evidence. The appeals court ruled that Ward's claims were angels dancing on the head of a pin. Glue sniffing was illegal. There was clear evidence of glue sniffing in the room. The judgment was affirmed.[17]

The case, however, pointed out new directions for the spread of the epidemic. Ward wasn't an adolescent, he wasn't a teenager, he wasn't a twenty-something. He was in his sixties. He was aiding and abetting a group of minors who were all female. Not one juvenile boy from a broken home was present at Duke's Motel. It was a signal that the contact trace map was still continuing to expand, but when the appellate decision appeared in June 1970, the map had been largely abandoned.

On December 30, 1970, Congress passed the Poison Prevention Packaging Act, which, like other pieces of federal legislation over the preceding decade, seemed to present legislators with a real chance to impose sanctions on model airplane glue as Congresswoman Mink had desired. The law required warnings on packages for all products deemed hazardous substances, all poisons, insecticides and fungicides, many foods, drugs, and cosmetics, and heating, cooking, and refrigerating fuels used in the home. It also required special packaging on products that could be harmful to children, thereby making it harder for them to open containers and accidentally hurt themselves. And accidental use was the thing. The Poison Prevention Packaging Act, like its predecessors, sought to protect people from products that could harm them through accidental error or common household use. It was not intended to fend off intentional misuse of household products, acts that Congress deemed outside of companies' control. After all, people adversely affected by products in their homes were victims. People who used those products for their own nefarious ends were not. Even though such thinking flew in the face of all of the previous decade's research, and particularly the research that developed after the 1967 Denver glue summit, Congress continued to cling to it for dear life, pushing model airplane glue to the margins and emphasizing that corporations were responsible for the things they could control.[18]

But this didn't mean that glue sniffing simply disappeared. In 1971, Columbia University's School of Public Health conducted a survey of American teenagers to gauge how many of them smoked marijuana. It was clearly an extension of assuming that the drug epidemic had morphed into something else, a reflexive response to the counterculture. The study showed that 15 percent of American teenagers had smoked pot, but it also discovered that 5 percent had sniffed glue. A figure of 5 percent was more than statistically significant, but Columbia admitted that the survey wasn't complete—it largely omitted children with "low educational attainment," just the sort of teenager who a decade of research had demonstrated was most likely to sniff glue.[19]

A year later, another letter to Ann Landers worried about a woman's stepchildren, one of whom had had two abortions by the age of sixteen and another who "is dropping acid and sniffing glue." The woman called them monsters. Landers called them pathetic. That was another consequence of the decline of glue's "epidemic" status.[20] No longer were children considered victims of a plague beyond their control. No longer was model airplane glue a scourge on America's youth. Now that the hyperbole had died away and the focus of people's attention was on other drugs, children who used such substances were no longer sad or sympathetic figures. They were pathetic monsters. The language that had once applied to the broader phenomenon was now being used to describe the users themselves.

There were, however, plenty of pathetic monsters still wandering the streets of the United States. In 1971, 3,397 cases were referred to the Maryland juvenile court, though only about one-third of those cases ever reached a judge. State intake aids handled the rest. Theft was a problem, as was vandalism. But along with drug offenses for marijuana, there were still plenty of glue-sniffing cases that either stood alone or stood as the fuel for robberies and other crimes. Back in the epicenter of the epidemic, Denver counselors and family therapists still regularly acknowledged cases of glue sniffing that hurt the families who sought their help.[21]

But boys outside the confines of Denver weren't immune, either, from model airplane glue or from other household solvents that worried so many researchers and activists. On June 26, 1972, Michael Harless, a fourteen-year-old Florida boy, was watching cartoons with his friend when they decided to try a new phenomenon they had learned just two days earlier. They took a toilet paper roll and stuffed the tissue into the center of the cardboard tube. They retrieved Michael's mother's can of Pam from the kitchen. Pam was a cooking spray designed to keep food from sticking to pots and pans, but it was also a Freon propellant. When the boys sprayed Pam into one side of the toilet paper tube and inhaled through the tissue filter from the other side, they got high. It was just another example of the dangerous use of household solvents, and this one ultimately killed young Michael. His mother's method of fighting for her son was at the same time obvious and bafflingly unique. She sued Pam.[22]

Companies like Testor had been largely immune from such litigation because from the outset, the company (and the HIAA) claimed to be horrified by the problem. They made films warning of the dangers of model airplane glue. They participated in the various sniffing conferences. They worked with lawmakers and researchers. But with the hysteria about household solvents gone by the 1970s, Pam's public image wasn't so damaged that American Home Products, the company that made Pam, had to do such things. Instead, they put a warning on the can that urged users to "avoid direct inhalation of concentrated vapors. Keep out of reach of children."[23]

In this case, however, the hysteria did not match the phenomenal seriousness of the actual problem. American Home Products knew of at least forty-five deaths occurring from the misuse of Pam between 1968 and 1972, so in March of that year, they changed the label. "CAUTION: use only as directed, intentional misuse by deliberately concentrating and inhaling the contents can be fatal." What they didn't do was recall the cans with the older warning. Michael's can was one of those. So his mother sued. A federal circuit court in Florida ruled for the company, pushing back against the new personal injury litigiousness that seemed to be sweeping the country and arguing that there was still a label that warned about the dangers of such behavior on the can. An appellate court, however, reversed the decision, making the case that the company knew about forty-five deaths from the inhalation of Pam, and thus the stronger warning was needed.[24] It was a glaring reminder that, although model airplane glue had been made undesirable through the addition of mustard oil, other products were still available and just as deadly. Pam paid the settlement, but the new label ensured that they wouldn't be sued again. They didn't make a cautionary film. They didn't call for a conference on the product's use. They didn't have to. At least forty-six people had died from using Pam, but the press hadn't created a campaign against it. And so it wasn't an epidemic.

In 1973, the Canadian Commission of Inquiry into the Non-medical Use of Drugs produced its final report. It noted that both the Hazardous Products Act and the Hazardous Products (Hazardous Substances) Regulations required special labeling on model airplane glue, nail polish remover, and other potentially dangerous intoxicants, but it didn't make them illegal. "Consequently, all of the above mentioned substances are legally available to anyone regardless of age or condition." The report noted that, while Testor's model airplane glue "was the most popular volatile solvent in Canada prior to 1968," after the addition of allylisothiocyanate, users began buying its competitors' products instead. The report was significant for several reasons. First, it belied Testor's claims that the company's bottom line wasn't affected by the loss of glue-sniffing customers. Second, it acknowledged that volatile products and toxic solvents were still readily available throughout Canada. And third, it admitted that the abuse of model airplane glue was far from over. "While paint thinner is the most popular solvent in Japan and in the Scandinavian countries, in North America airplane glue and nail polish remover" were the solvents "most often used for psycho-tropic purposes." They were cheap, available, and used in earnest by Canadian children from coast to coast.[25]

It was not surprising that the United States ignored the report. The United States has a long-standing tradition of ignoring its neighbor to the north. Besides, the country had moved on to worry about other drugs. The harangues of officials in both countries continued to fall not on deaf ears but on ears that were already otherwise engaged. Their efforts were drowned out by the larger imperatives of Vietnam, the burgeoning Black Power movement, and the drug-laden counterculture that seemed to show no signs of abating. It was the same sort of dulled momentum that kept the Deep South from considering glue-sniffing legislation until the second half of the 1960s (see chapter 8), and which now had enveloped the country.

The drug crackdown and consolidation that developed in the 1960s, largely as a response to the counterculture, never included model airplane glue, but the hysteria that developed around glue sniffing would set a precedent for the modern War on Drugs, a new moralistic assault on drug use that would begin with the presidency of Richard Nixon in 1969. Nixon signed the Controlled Substances Act of 1970, which replaced the Harrison Narcotics Act of 1914 as the principal federal method to imprison drug users and dealers. In 1973 Congress created the Drug Enforcement Agency as the principal police force for narcotics. Nixon referred vaguely to a "total war" on "an enemy with many faces," which was generally interpreted to mean those opposed to his interpretation of "law and order," such as civil rights activists, members of the counterculture, and drug users.[26] He was, essentially, out to create a sense of paranoia among those who were prone to fearing enemies with many faces. The rhetoric established "law and order" as the true American good, which would thereby allow adherents to posit anything against that vague standard as being somehow frightening or threatening. Certainly there are parallels to the paranoia over drugs created by the Nixon administration and the paranoia over pot that led to the Marihuana Tax Act of 1937 (see introduction), but the most immediate precedent on which the administration could draw was the hysteria that developed over model airplane glue. It was a coordinated version of the parent-, scientist-, and media-driven epidemic of the 1960s.[27]

Although the paranoia over drug abuse diminished somewhat during the Carter administration (with the president even suggesting the possible decriminalization of marijuana), it expanded to record heights during the administration of Ronald Reagan. The discovery of AIDS and the overblown race-infused hysteria over crack cocaine allowed drug warriors to take advantage of new public fears. The Comprehensive Crime Control Act of 1984, the Anti–Drug Abuse Act of 1986, and the Anti–Drug Abuse Act of 1988 created myriad new controls and stiffer penalties for drug users and dealers. During its near unanimous assent to the 1988 legislation, Congress declared that "the legalization of illegal drugs," in any form, would constitute an "unconscionable surrender in a war in which . . . there can be no substitute for total victory."[28] Such talk only emphasized punishment, precluded adequate resources for treatment, and grew the prison system to unprecedented size. It also unfairly targeted lower-class black males, generating what Michelle Alexander has proven to be a permanent underclass that has, even in the new century, only furthered the racial divide on the backs of millions of nonviolent criminals.[29] It was the kind of reaction that could only develop while riding the wave of a monumental moral panic, which itself rode the wave of the panic over model airplane glue that came immediately before it, rising as it did in tandem with the budding counterculture.[30]

But it was more than just panic. Not only did the hysteria over model airplane glue set the parameters of later drug wars, it seems, at least anecdotally, to have actively provided fodder for them. In 1965, in the heart of the glue-sniffing epidemic, juvenile drug arrests for substances other than model airplane glue were rare. By 1970 this was no longer the case. Officials arrested children under the age of seventeen for marijuana, cocaine, heroin, and synthetic drug crimes at exponentially greater rates. That five-year jump is the only such explosion in the decades that followed. The kids, it seems, were graduating from glue.

Still, just because the headlines had gotten smaller and the news stories had moved farther back in the paper, this didn't mean that the problem had magically been solved. In 1974 Washington, DC, area churches began locking their doors to protect against vandals and other delinquents who snuck into the chapels at night to sniff glue or smoke pot. At least six Maryland residents had died from sniffing Freon that year. In 1975 a Fairfax, Virginia, girl died suddenly after inhaling the fumes of an aerosol deodorant. In 1976 an Owensboro, Kentucky, teenager with a long history of glue sniffing was arrested for murdering a 7-Eleven night manager.[31] Toxic solvents were no longer the subject of national fear and hyperbole, but they could still make their presence felt.

A year after Owensboro, the National Institute on Drug Abuse (NIDA) published a review of inhalants. Although the glue-sniffing epidemic had largely run its course, the NIDA was clear: the use of toxic solvents was still a daily problem in the United States and throughout the world. "National survey figures for inhalant abuse report levels of abuse roughly comparable to that for the major hallucinogens such as LSD and for the stimulant cocaine," and those figures, the NIDA assumed, were low. The problem was probably far more widespread. Some solvents were more toxic than others, some had more animal research than human research. Toluene was still being used as an active ingredient in many household products, but there were scores of other substances that could now be added to the list. The NIDA pressed for more sustained research, more clinical studies. Almost a decade had passed since the final throes of the glue-sniffing epidemic, but because the

## Juvenile Drug Arrest Rates by Race, 1965–1985 (per 100,000 population)

|                    | 1965 | 1970  | 1975  | 1980  | 1985  |
|--------------------|------|-------|-------|-------|-------|
| **MARIJUANA**      |      |       |       |       |       |
| White              | 6.9  | 181.8 | 436.1 | 393.5 | 284.7 |
| Black              | 31.0 | 100.8 | 312.8 | 379.8 | 378.2 |
| Other              | 18.2 | 163.4 | 246.5 | 128.0 | 160.0 |
| **HEROIN/COCAINE** |      |       |       |       |       |
| White              | 1.3  | 18.2  | 14.1  | 19.4  | 42.4  |
| Black              | 4.0  | 53.0  | 36.4  | 31.0  | 120.8 |
| Other              | 6.8  | 11.3  | 21.3  | 2.8   | 7.2   |
| **SYNTHETIC DRUGS**|      |       |       |       |       |
| White              | 1.7  | 44.8  | 14.4  | 11.6  | 7.7   |
| Black              | 2.8  | 54.4  | 7.0   | 4.9   | 7.7   |
| Other              | 9.9  | 47.7  | 8.0   | 6.0   | 1.7   |

JUVENILE DRUG ARREST RATES: Juvenile in this context indicates children ages 10-17. Numbers given signify arrests per 100,000 juveniles. Source: Federal Bureau of Investigation, *Crime in the United States: Uniform Crime Reports, 1991* (Washington, DC: Uniform Crime Reporting Program, 1992), 286-87.

hysteria had dissipated, so too had the coordinated medical effort to solve the problem. While there were plenty of studies from the 1970s examining the health effects of aerosols and other products, much of the literature used by the NIDA came from the 1960s, when the scientific community had rallied around one central issue.[32] It was a dense document, relatively unheralded and largely unread by the vast majority of the American public. The problem of household solvents still affected children and families every day, but without the hysteria that accompanied the glue-sniffing epidemic, no one was rushing out to consume the new data that might provide them the keys to a troubled kingdom.

Other sources, however, were far less dense, far more easily consumed. Douglas Glenn Colvin had grown up in Queens in the 1960s, right in the thick of one of the principal areas of infection during the glue-sniffing epidemic. In the mid-1970s, he and a group of friends from Queens formed a new band, The Ramones, which would be principally responsible for a new American "epidemic": punk rock. The band members all changed their names, Colvin becoming Dee Dee Ramone before the group's eponymous debut album appeared in April 1976, just months before the Owensboro murder. Dee Dee played bass and was the principal songwriter for the band. One of his compositions, the sixth song on the first side of the album, was an homage to his Queens childhood in the 1960s. "Now I Wanna Sniff Some Glue" was only 1:34 minutes long, the length itself a statement on the short attention spans of youth. "Now I wanna sniff some glue," wailed Joey Ramone, the band's lead singer. "Now I wanna have somethin' to do. All the kids wanna sniff some glue, all the kids want somethin' to do."[33]

All the kids of the 1960s had at least a passing, peripheral experience with the nation's crazed hysteria over the glue-sniffing epidemic. Many of them had actually participated in the dangerous fad that killed well more than one hundred children and tormented the families of tens of thousands of others. By 1976, however, although there were still scattered instances of problems with inhalants and household solvents of all kinds, although researchers were still inherently unclear about the long-term medical consequences of such use, and although law enforcement officials still had to remain vigilant to prevent more cases like Fairfax and Owensboro, the kids had largely moved on from model airplane glue and its dangerous antecedents. And the kids were all right.

# Conclusion

THE INFLUENZA EPIDEMIC OF 1918 and 1919 killed between 20 and 40 million people worldwide. It infected 28 percent of all Americans and killed more than 675,000 people in the country. The average American life span dropped by ten years because of the plague. Although it probably originated in China, the affliction was known as the Spanish flu because of the early death toll emanating from the Iberian Peninsula. It first appeared in the United States in Kansas military camps, but the real scourge emanated from the port of Boston in September 1918. By the end of October, more than 200,000 were dead. Public health departments issued masks for people to wear in public. There was a massive coffin shortage throughout the country. Ultimately, the frightened populace turned to science, its germ theory, antiseptic surgery, and an eventual vaccine to end the horror of death that had swept the nation and the world.[1]

As terrible as it was, such was what constituted an actual medical epidemic. The illness spread quickly. Public information campaigns tried to keep people safe and limit the infection rate and death toll until adequate vaccines could stop it in its tracks. When the vaccine appeared and people began getting the shot, the epidemic was over.

The glue-sniffing epidemic of the 1960s had some similar traits to what might be called a traditional epidemic. The illness spread quickly. Public information campaigns tried to keep people safe and limit the infection rate. But this is where the similarities stopped. The illness was global, but it obviously stopped well before the shores of a demographic catastrophe like the early-century influenza epidemic. It also spread quickly precisely because the public information campaign existed. Those infected infected themselves willingly, and when they discovered through the media that there was an easy method of infection in their dresser drawers, they actively took part to see what all the fuss was about. The epidemic didn't end because a suitable vaccine appeared. It ended because those running the information campaign stopped informing people. That is because, for all of the medical discussions, the glue-sniffing epidemic was not medical. It was, at its heart, semantic.

Perhaps this isn't fair. The Testor Corporation did eventually discover an adequate irritant to put in its products and did willingly share that information with its competitors. There did seem to be a legitimate "fad" nature to the use of model airplane glue. And laws that punished either sniffers or sellers or both created real consequences for users that certainly curtailed the willingness of many to start or continue their habits. Juvenile courts, detention facilities, and counselors also worked diligently to help children who needed it. Researchers experimented extensively to understand the nature and circumference of the problem and to devise solutions to the growing menace. It was the kind of all-fronts comprehensive assault that medical and governmental authorities used to combat the influenza epidemic a half-century prior.

Still, it was hardly a traditional epidemic. It was a disease spread by the media, even though this was never the media's intent. The notion of the problem's status as "epidemic" was in fact created by the media. In this sense, the glue-sniffing epidemic of the 1960s was, more than anything else, a cultural construct, a self-fulfilling prophecy that witnessed initial cries of doom, which only fed the potential for doom across the country. It was a moral panic with a broad confluence of parents from all sectors—politics, law enforcement, science, academia, the media—as generative moral agents. Unlike the scourge of 1918, the epidemic didn't stop when people stopped dying. People kept dying. But by that time, there were other threats to the social fabric of the country, and media outlets shifted their focus to them, leaving household solvents behind even though children still used them regularly. One of the other principal differences between the glue-sniffing epidemic and traditional epidemics was that only a certain segment of the population was vulnerable. "Youth," as sociologist Kenneth Thompson has noted, "may be regarded as both at risk and a source of risk in many moral panics."[2]

The official profile of the potential glue sniffer—a poor, withdrawn boy from a broken home with a low intelligence quotient and poor grades in school—didn't always hold. There were myriad examples of girls sniffing glue, though they were still far less likely than boys to adopt the habit. There were also myriad examples of users in their late teens, users in full adulthood, all eager to find the same easy high that boys were trying to find. In addition, those adolescent boys who did sniff glue weren't always poor, weren't always withdrawn, and weren't always from a broken home. Still, even if officials had failed in creating a comprehensive profile, they had certainly pegged the majority of users. Middle-class, propitious adults weren't sniffing glue. More importantly, they weren't in real danger of ever sniffing glue. The epidemic could theoretically affect anyone, but in reality only a segment of the population was vulnerable.

And so the epidemic hovered as a broad idea clouding the sky above the American public. "Epidemic" in 1918 was a definition. In the 1960s, it was a label.

Regardless of whether the problem was a proper epidemic or not, however, it was clear that the furor really had subsided by the early 1970s. The media might have been responsible for spreading the idea of glue sniffing, but it was also diligent about plotting the trajectory of that spread. As the incidences of crazed glue-induced behavior subsided, the media found itself with nothing to report. When Testor products became less useful to glue sniffers, the media covered the irritant addition as a culmination of the problem, as the vaccine for which everyone had waited. And whether or not it was a traditional epidemic, the media wasn't fabricating reports of glue sniffing. The fact remained: Beginning in 1959, the American people discovered that adolescents were inhaling model airplane glue and other solvents for the purpose of getting high. As the reports of this behavior spread, more and more examples of such behavior began appearing. Over the course of several years, tens of thousands of glue sniffers were caught, and even more participated without getting caught. States and municipalities across the country responded with legislation to curb the problem. Researchers devoted time and treasure to understanding the legal and medical ramifications of what the aberrant behavior meant. The hysteria that resulted was less because children were doing drugs and more because children were improperly using products that were designed to be in their hands.

It would be tempting, then, to describe the nation's hand-wringing as a form of collective guilt—adults frustrated and flagellating themselves for having encouraged their children to engage in hobbies with such sinister ancillary effects. That, however, would be a label similar to "epidemic," and it wouldn't be helpful in explaining just why the hysteria came and went as it did. But make no mistake, the epidemic existed precisely because of that hysteria. With the long history of sniffing gasoline and other products well before the 1960s firmly in place, the likelihood that the first group of adolescents to try inhaling model airplane glue was the same group arrested in

Colorado in 1959 is absurd. The practice of glue sniffing began well before the epidemic, and it continued well after the epidemic had run its course. In between, there existed a call-and-response between newspapers, researchers, principals, parents, and kids, the first four decrying the practice and working to find solutions, the last learning that the models waiting for them on their bookshelves held hidden secrets.

The final element contributing to the erosion of the glue-sniffing epidemic was the birth of the student movement and its attendant countercultural values. In the 1960s the broad-based use of marijuana and psychotropic drugs gave parents and the media a new scourge to pursue, thus moving the focus from model airplane glue. In the long run, for the rest of the twentieth century, the growing use of such drugs served to normalize the act. Drug use remained a problem, something to legislate, something for moralists to use as another sign that society was in decay, but at the same time the use of those drugs was no longer surprising. This functional loss of public innocence wasn't simply a creature of drug use. The 1960s were a decade of civil rights unrest, Vietnam, and several high-profile assassinations. Drug use was a menace, of course. It was a criminal act. And ultimately the wrangling over model airplane glue and the broader tumult of the counterculture and its discontents would lead to the overblown American War on Drugs. But at the same time, as the sixties became the seventies, the seventies became the eighties, and the century moved through its final phases, no one was shocked that drugs were being used.

That dichotomy fed from a growing liberalization of American thinking about drugs—through, if nothing else, constant exposure—and an expanding government more beholden to corporate interest and reactionary politics. It created what Noam Chomsky has called a "democracy deficit," a widening gap between the public will and the actions of a government increasingly isolated from that public. And so Richard Nixon expanded the federal drug budget and created new agencies to police drugs. The Reagan administration passed two new major pieces of legislation in 1986 and 1988 and an even larger budget for the drug war. Michelle Alexander has effectively demonstrated that the government used such tactics to create a replacement system for Jim Crow segregation, giving itself the permanent racial underclass that has always existed in the United States. Still, this is not to say that the country could never be shocked by drugs or that the government couldn't be sincere in its concern. Craig Reinarman and Harry G. Levine have demonstrated that the crack scare of the 1980s, though racial in its scope, policing, and punishment, created yet another moral panic in the country—complete with overwrought jeremiads about the state of morals in America and conspiracy theories from those groups most affected that the government was somehow to blame.³ In the face of those new menaces, and with a government poised to use them to reinstitute a modern version of the turn-of-the-century convict lease system, model airplane glue faded from public perception.⁴

The phenomenon was similar to the sexual revolution which attended that same countercultural movement. Much went into creating the sexual revolution. The work of Alfred Kinsey and Masters and Johnson. The censorship cases stemming from Grove Press publications. The birth of the sexual advice manual. The advent of birth control and new treatments for venereal disease. The resulting promiscuity that grew from such sources was a horrible travesty to many American moralists, yet another sign that the country had gone the way of Sodom and Gomorrah. Ultimately, however, the sexual revolution normalized promiscuity, and sexual activity prior to (or without the cover of) marriage became commonplace. Plenty of moralists derided the practice. Many still do. And AIDS in the 1980s created a moral panic all its own. But no one any longer claimed with any statistical justification that sexual promiscuity was aberrant behavior.

So it went for American drug use. The War on Drugs that ultimately became the proposed vaccine for that particular epidemic is still ongoing. There are more Americans in jail for nonviolent drug offenses than for any other crime. The country spends countless amounts of blood and treasure trying to police the problem. At the same time, however, any admission that someone smokes pot or has experimented with drugs at some point in their lives is hardly shocking. When Lenny Bruce performed in the 1950s and 1960s, doing bits about model airplane glue and other drugs, he shocked a nation and eventually ended up in court for his efforts. Today, comedians regularly talk about their own drug use to laughter and supportive applause.

Such normalizing practices, of course, have happened for better and for worse. There are still regular reports of household solvent abuse by teens and adolescents across the country. New threats present themselves all the time. The normalization of America's relationship with drugs, however, ensures that though there will be plenty of local worry when such problems arise, the mass national hysteria that accompanied the realization that children were sniffing model airplane glue will remain rare. It will always cause concern when adolescents are discovered, for example, huffing gasoline, but with the mass death that comes attendant with drug cartels, organized crime, and other harder drug sources, the problem will always seem to those not specifically experiencing it a modest indiscretion at best.

Thus the epidemic subsided. Not because the adolescent abuse of household solvents stopped, but because by the end of the 1960s, there were so many examples of things that could be worse. The media had moved on. Parents had new things to fear. Still, for one brief decade, adolescents found themselves in the center of a national firestorm—the cause of concern and the signposts for all that had devolved in immoral America. For one brief decade, they were the Louis Pasteurs of junkiedom.

# Appendix One

**A Description of Active Chemicals in Organic Solvents From the National Institute on Drug Abuse's *Research Monograph 15, Review of Inhalants: Euphoria to Dysfunction***

*The NIDA report, published in 1977, included descriptions of the most problematic of the chemicals in organic solvents, particularly those that were intentionally inhaled for intoxicating effect. Because of its late publication date, the information was general, complete, and without the hysteria of reports in the 1960s. Its descriptions, then, provide a simple glossary to some of the more damaging substances being abused during the glue-sniffing epidemic.*[1]

## Contents

| | |
|---|---|
| n-Hexane | 179 |
| Toluene | 180 |
| Gasoline | 180 |
| Aerosols | 180 |
| Benzene | 180 |
| Xylene | 181 |
| Naphthalene | 181 |
| Chlorinated Hydrocarbons | 181 |
| Aliphatic Hydrocarbons | 181 |

<u>n-Hexane</u>
n-Hexane exposure by inhalation clearly is associated with polyneuropathy, which is predominantly motor. A latent period of 6 to 10 weeks is usually necessary but months to years may elapse between initial exposure and clinical effects following lower levels of exposure. The question of persistent cerebral dysfunction following n-hexane polyneuropathy has not been addressed. Injury to other organ systems has not been identified.[2]

## Toluene

Toluene ($C_6H_5CH_3$; methylbenzene, toluol, phenylmethane) is a substance preferred by many inhalant users. Commercial products containing toluene are sought after and used for long periods of time. Since commercial products containing toluene usually contain a wide variety of other volatile organics, generalizations from single case reports have precarious validity. In contrast with n-hexane there is no single predominating target organ system that shows a response to toluene. The diversity of responses associated with toluene suggests that other substances, either alone or in combination, are the primary toxic agents. Toluene users who do not develop significant injury are grossly underrepresented in the literature.... Central and peripheral nervous system, liver, and kidney injury have occurred in association with toluene use.[3]

## Gasoline

Gasoline (petrol) vapor inhalation occurs primarily among younger children or in isolated cultures where a very limited variety of volatile substances is available. Various gasoline additives present special hazards. Triorthocresyl phosphate (TCP) is an established cause of both upper and lower motor neuronal degeneration with spastic muscle-wasting disorders. Benzene, a common ingredient, is an established cause of subacute and chronic disorders of the hematopoietic system, including various combinations of cytopenia and delayed occurrence of leukemia. Organic lead additives may cause acute and chronic lead encephalopathy. The diversity of clinical effects reported in gasoline inhalers is consistent with the effects of these various additives.[4]

## Aerosols

Abuse of Freon-pressurized aerosol products is common. (Not all aerosols contain Freons, and many contain solvents other than Freon.) The great variety of clinical manifestations attributed to these products is not surprising because of the diversity of contents. The consistently recognized syndrome is sudden death associated with vigorous exertion immediately after inhaling Freon. Myocardial sensitization to endogenous epinephrine with ventricular fibrillation is one accepted mechanism. A number of case reports and reviews of Freon use appear in the current literature.[5]

## Benzene

Benzene ($C_6H_6$; benzol, phenyl hydride, cyclohexatriene) is a volatile, highly flammable liquid with a characteristic odor. It is only slightly soluble in water, but freely soluble in alcohols and other organic solvents. Benzene is obtained as a by-product from petroleum and coke oven emissions. Benzene is used as a substrate in the manufacture of many aromatic compounds and as a solvent for waxes, resins, plastics, lacquers, varnishes, and paints. The use of benzene as a solvent has been more limited in recent years, due to its recognized myelotosic potential. Benzene is nevertheless often present in varying quantities in hydrocarbon solvent mixtures, including gasoline ... and assorted thinners and solvents.[6]

### Xylene
Xylene ($C_6H_4(CH_3)_2$; xylol, dimethyl benzene) is a volatile, flammable liquid at room temperature. It is practically insoluble in water, but freely miscible with most organic liquids.... Xylene is produced from both petroleum and coal tar and is used as a solvent or filler in a myriad of commercial products including paints, lacquers, varnishes, dyes, inks, cements, cleaning fluids, gums and resins, oils, rubber, and gasoline. Xylene is also used in the chemical industry as a synthetic intermediate. Because of such widespread use and availability, there is a potential for abuse of xylene.[7]

### Naphthalene
Naphthalene ($C_{10}H_8$; naphthalin, tar camphor) is a white crystalline solid that volatilizes appreciably at room temperature. It is obtained primarily from coal tar and petroleum. Naphthalene is insoluble in water, but quite soluble in organic solvents. It is used commercially as a substrate for synthesis of a number of chemicals, as a moth repellent, as a toilet bowl deodorant, and as a veterinary antiseptic and vermicide. Naphthalene is also a common constituent of such hydrocarbon mixtures as gasolines, thinners, and assorted organic solvents.[8]

### Chlorinated Hydrocarbons
Among the chlorinated hydrocarbon solvents there is a potential for injury of various organ systems, especially neuropathy and liver and kidney injury. These substances also sensitize the myocardium to epinephrine-induced dysrhythmia.[9]

### Aliphatic Hydrocarbons
The aliphatic hydrocarbons (paraffins) are a series of straight- and branched-chain hydrocarbons, including alkanes (saturated), alkenes (one double bond), alkadienes (two double bonds), and alkatrienes (three double bonds). Those aliphatics with at least one double bond are also known as olefins. These hydrocarbons occur naturally as mixtures in petroleum, from

which they may be separated by cracking processes and fractional distillation. For purposes of the present paper, discussion will be limited largely to the straight-chain alkanes of intermediate length. Alkanes containing fewer than five carbon atoms are gases at room temperature, while pentane and the higher alkanes are volatile, flammable liquids. All are insoluble in water, but miscible in other organic solvents. These aliphatics are used individually as solvents and are major constituents of such mixtures as petroleum ether, gasoline, kerosene, and assorted thinners and solvents.[10]

# Appendix Two

## A Documentary Account of Glue-Sniffing Legislation in the Deep South

*The Deep South's laws ran the gamut from simple misdemeanor violations for sniffing to treating model airplane glue as a Schedule One narcotic. The laws are presented here to provide an easily comparable account of those laws and how legislators went about forming them. Text in brackets ([ ]) was excised from the original draft through legislative amendment. Text in braces ({ }) was added to the original draft through legislative amendment. Italicized language represents proposed additions that failed to pass.*

## Contents

| | |
|---|---|
| Louisiana | 183 |
| Florida | 187 |
| Mississippi | 189 |
| Georgia | 190 |
| Alabama | 193 |

### LOUISIANA

Act No. 110

House Bill No. 752

By: Messrs. Sapir, Chaisson, Cole, Johnston, Gill, Fortier, Triche, Boesch, Smither, Gremillion, Hessler, Bordes, S. M. Morgan, O'Hearn, Hillensbeck, Causey, Beeson, Richardson, Casey, Bel, LeBreton, Sullivan, O'Brien, Crais, Strother, Guidry, Keller, Early, Dwyer, Talbot, Marcel, Daley, Gregson, Lauricella, Coreil, Anzelmo, Hoffman, Keogh, Bernhard, Nunez, Miller, Simon, Vesich, Lacy, Smith and Mrs. Walker.

## AN ACT

To amend Title 14 of the Louisiana Revised Statutes of 1950 by adding thereto a new Section to be designated as Section 93.1 thereof, to define the term, "model glue," to provide for the sale and use thereof and to provide penalties in connection therewith.

Be it enacted by the Legislature of Louisiana:

Section 1. Section 93.1 of Title 14 of the Louisiana Revised Statutes of 1950 is hereby enacted to read as follows:

§ 93.1. Model glue; use of; unlawful sales to minors; penalties

A. As used in this [S]section, the term "model glue" shall mean any glue or cement of the type commonly used in the building of model airplanes, boats and automobiles and which contains one or more of the following volatile solvents: (1) toluol, (2) hexane, (3) trichlorethylene, (4) acetone, (5) toluene, (6) ethyl acetate, (7) methyl ethyl ketone, (8) trichlorochthane, (9) isopropanol, (10) methyl isobutyl ketone, (11) methyl cellosolve acetate, (12) cyclohexanone, or (13) any other solvent, material, substance, chemical or combination thereof having the property of releasing toxic vapors.

B. It shall be unlawful for any person to intentionally smell or inhale the fumes of any type of model glue for the purpose of causing a condition of, or inducing symptoms of, intoxication, elation, euphoria, dizziness, excitement, irrational behavior, exhilaration, paralysis, stupefaction or dulling of the senses or nervous system;[,] or for the purpose of, in any manner, changing, distorting or disturbing the audio, visual or mental processes; provided, however, that this section shall not apply to the inhalation of any anesthesia for medical or dental purposes.

C. It shall be unlawful for any person to sell any type of model glue to a minor for any reason whatsoever.

D. It shall be unlawful for any person to sell or otherwise transfer possession of any type of model glue to any minor under the age of twenty-one[,] for any purpose whatsoever, unless the minor receiving possession of the model glue is the child or ward of and under the lawful custody of the vendor, donor or transferor of the glue.

E. Any person violating any provisions of this section shall be guilty of a misdemeanor and, upon conviction thereof, shall be fined not less than twenty-five dollars nor more than one hundred dollars or imprisoned for not more than ninety days for each such offense or both.

Section 2. All laws or parts of laws in conflict herewith are hereby repealed.

Approved by the Governor: July 13, 1966.

A true copy:
WADE O. MARTIN, JR.
 Secretary of State.

**FAILED BILL AMENDMENT FROM LOUISIANA SENATE** AN ACT

To amend Title 14 of the Louisiana Revised Statutes of 1950 by adding thereto a new Section to be designated as Section 93.1 thereof, to define the term, "model glue," to provide for the sale and use thereof and to provide penalties in connection therewith.

Be it enacted by the Legislature of Louisiana:

Section 1. Section 93.1 of Title 14 of the Louisiana Revised Statutes of 1950 is hereby enacted to read as follows:

§ 93.1. Model glue; use of; unlawful sales to minors; penalties

*A. As used in this Section, the term "model glue" shall mean any glue or cement of the type commonly used in the building of model airplanes, boats and automobiles, and which contains one or more of the following volatile solvents:*
*Acetone, (2) Amylacetate, (3) Benzol or benzene, (4) Butyl acetate, (5) Butyl alcohol, (6) Carbon tetrachloride, (7) Chloroform, (8) Cyclohexanone, (9) Ethanol or ethyl alcohol, (10) Ethyl acetate, (11) Hexane, (12) Isopropanol or isopropyl alcohol, (13) Isopropyl acetate, (14) Methyl "Cellosolve" acetate, (15) Methyl ethyl ketone, (16) Methyl isobutyl ketone, (17) Toluol or toluene, (18) Trichloroethylene, (19) Triscresyl phosphate, and (20) Xylol or xylene, or any other solvents, material, substance, chemical, or combination thereof, having the property of releasing toxic vapors.*

B. It shall be unlawful for any person to intentionally smell or inhale the fumes of any type of "model glue," or to induce any other person to do so, for the purpose of causing a condition of, or inducing symptoms of intoxication, elation, euphoria, dizziness, excitement, irrational behavior, exhilaration, paralysis, stupefaction or dulling of the senses or nervous system, or for the purpose of, in any manner, changing, distorting or disturbing the audio, visual or mental processes; provided, however, that this section shall not apply to the inhalation of any anesthesia for medical or dental purposes.

C. It shall be unlawful for any person to intentionally possess any type of "model glue" for the purpose of using the same in the manner prohibited by Sub-section B.

D. It shall be unlawful for any person to intentionally possess, buy, sell or otherwise transfer any type of "model glue" for the purpose of inducing or aiding any other person to violate the provisions of Sub-sections B or C.

E. It shall be unlawful for any person to knowingly and intentionally sell or otherwise transfer possession of any type of "model glue" to any minor under the age of eighteen, for any purpose whatsoever, unless such seller or transferror [sic] first receives the written consent of said minor's parent or guardian. Where such consent is received, the seller or transferror [sic] shall make a written record of such transaction, showing the name, address, sex and age of the minor, as well as the name and address of the consenting parent or guardian, which said record must be kept available for inspection by the Department of Police for a period of at least one year from date of transfer; provided that such records need not be kept if the seller or transferror [sic] retains the consent letters where such letters contain all of the required information, as herein set forth.

F. It shall be unlawful for any wholesaler of merchandise to sell or transfer any quantity of "model glue" to any person whatsoever, who is not, or does not represent a recognized bona fide retail dealer in merchandise customarily handling such product in the ordinary course of his or its business at a fixed location. If any such prohibited sale is made by any wholesaler, it shall constitute prima facie evidence of an intent to violate the provisions of Sub-sections B and C.

G. It shall be unlawful for any person, firm or corporation to engage in the retail sale of "model glue" unless such person, firm or corporation is a recognized bona fide retail dealer (at a fixed location) in merchandise, customarily handling such product in the ordinary course of his or its business.

H. It shall be unlawful for any bona fide recognized retail dealer, as defined in Sub-section G, to sell more than one tube, or other minimum sized container, of "model glue" to any one customer within any 24 hour period; provided, that the provisions of this Sub-section do not apply to the sale of "model glue" together with a "hobby" or "model" kit.

I. It shall be unlawful for any bona fide recognized retail dealer, as defined in Sub-section G, to keep, maintain or prominently display in his business establishment any stock or supply of "model glue" in such manner or place as to make the same accessible to customers, or other members of the public, without actually receiving the same directly from the retailer, or his bona fide authorized representative.

J. Any person violating any provisions of this section shall be guilty of a misdemeanor and, upon conviction thereof, shall be fined not less than twenty-five dollars nor more than one hundred dollars or imprisoned for not more than ninety days for each such offense or both.

Section 2. All laws or parts of laws in conflict herewith are hereby repealed.

## FLORIDA

CHAPTER 67-416

Senate Bill No. 893

AN ACT relating to model glue; prohibiting inhalation; providing a penalty; providing an effective date.

*Be It Enacted by the Legislature of the State of Florida:*

Section 1.
(A) Definition—As used in this Chapter, the term "model glue" {shall not mean any of the below set forth ingredients or types when used industrially and in the usual course of business, but} shall mean any glue or cement of the type commonly used in the building of model airplanes, boats and automobiles and which contains one or more of the following volatile solvents:

(1.) Acetone
(2.) Amylacetate
(3.) Benzol or benzene
(4.) Butyl acetate
(5.) Butyl alcohol
(6.) Carbon tetrachloride
(7.) Chloroform
(8.) Cyclohexanone
(9.) Ethanol or ethyl alcohol

(10.) Ethyl acetate
(11.) Hexane
(12.) Isopropanol or isopropyl alcohol
(13.) Isopropyl acetate
(14.) Methyl "Cellosolve" acetate
(15.) Methyl ethyl ketone
(16.) Methyl isobutyl ketone
(17.) Toluol or toluene
(18.) Trichloroethylene
(19.) Tricresyl phosphate
(20.) Xylol or sylene

or any other solvent, material, substance, chemical or combination thereof, having the property of releasing toxic vapors.

(B) *Inhalation prohibited*—It shall be unlawful for any person to intentionally smell or inhale the fumes of any type of "model glue" as defined in this Chapter, or to induce any other person to do so, for the purpose of causing a condition of, or inducing symptoms of intoxication, elation, euphoria, dizziness, excitement, irrational behavior, exhilaration, paralysis, stupefaction or dulling of the senses of the nervous system, or for the purpose of, in any manner, changing, distorting or disturbing the audio, visual or mental processes; provided, however, that this section shall not apply to the inhalation of any anesthesia for medical or dental purposes.

(C) *Possession regulated*—It shall be unlawful for any person to intentionally possess any of "model glue" for the purpose of using the same in the manner prohibited by sub-section (B) of this section.

(D) *Inducing other persons*—It shall be unlawful for any person to intentionally possess, buy, sell or otherwise transfer any type of "model glue" for the purpose of inducing or aiding any other person to violate the provisions of sub-section (B) of this section.

Section 2.—*Penalties*—Any person who violates any of the provisions of this Chapter shall upon conviction be guilty of a misdemeanor, punishable by a fine of not more than $500.00, or by imprisonment in the county jail for a term of not more than six months, or by both such fine and imprisonment in the discretion of the court.

[Section 3.—Retail Sale—(A)

(B) It shall be unlawful for any bona fide recognized retail dealer, as defined in subsection (A) of this section to sell more than one tube, or other minimum sized container, of "model glue" to any one customer within any 24 hour period; provided, that the provisions of this subsection do not apply to the sale of "model glue" that is packaged with a "hobby" or "model" kit.

(C) It shall be unlawful for any bona fide recognized retail dealer, as defined in subsection (A) of this section, to keep, maintain or prominently display in his business establishment and stock or supply of "model glue" in such a manner or place as to make the same accessible to customers or other members of the public, without actually receiving the same directly from the retailer, or his bona fide authorized representative.]

Section 3.—*Effective Date*—This act shall take effect September 1, 1967.

Approved by the Governor July 26, 1967.
Filed in Office Secretary of State July 27, 1967.

## MISSISSIPPI

Chapter 347

House Bill No. 281

AN ACT to prevent the intentional inhaling of such glues or cements whose fumes disturb in any manner any part of the nervous system; to prevent the selling or giving of such glues and cements to any minor; and for related purposes.

*Be it enacted by the Legislature of the State of Mississippi:*

Section 1. It shall be unlawful for any person to intentionally smell or inhale the fumes of any type of model glue for the purpose of causing a condition of, or inducing symptoms of, intoxication, elation, euphoria, dizziness, excitement, irrational behavior, exhilaration, paralysis, stupefaction, or dulling of the senses or nervous system; or for the purpose of, in any manner, changing, distorting or disturbing the audio, visual or mental processes.

Section 2. As used in this act, the term "model glue" shall mean any glue or cement of the type commonly used in the building of model airplanes, boats and automobiles and which contains one or more of the following volatile solvents: (1) toluol, (2) hexane, (3) trichlorethylene, (4) acetone, (5) toluene, (6) ethyl acetate, (7) methyl ethyl ketone, (8) trichlorochthane, (9) isopropanol, (10) methyl isobutyl ketone, (11) methyl cellosolve acetate, (12) cyclohexanone, or (13) any other solvent, material, substance, chemical or combination thereof having the property of releasing toxic vapors.

Section 3. It shall be unlawful for any person to sell or otherwise transfer possession of any type of model glue to any minor for any purpose whatsoever, unless the minor receiving possession of the model glue is the child or ward of and under the lawful custody of such person.

Section 4. Any person violating any provision of this act shall be guilty of a misdemeanor and, upon conviction thereof, shall be fined not [less than One Hundred Dollars ($100.00) nor] more than Five Hundred Dollars ($500.00), or imprisoned for not more than ninety (90) days, or both, for each such offense.

Section 5. This act shall take effect and be in force thirty (30) days from and after its passage.

Approved July 12, 1968.

## GEORGIA

Intentional Inhaling of Fumes of Model Glue, Etc.

No. 1122 (Senate Bill No. 205).

AN ACT to prohibit the intentional inhaling or smelling of fumes from model glue for the purpose of causing a condition of intoxication, stupefaction, euphoria, excitement, exhilaration or dulling of the senses or nervous system; to define "model glue"; to prohibit the intentional possession, buying, selling or transferring of possession or receiving possession of model glue for the purpose of violating or aiding another person to violate the provisions of this Act; to provide that model glue shall not be sold or transferred to persons under the age of eighteen (18) [twenty-one (21)] without the written consent from said person's parent or guardian;

to provide for the keeping of records of sales of model glue to persons under the age of eighteen (18) [twenty-one (21)]; to provide that violation of this Act shall be a misdemeanor and punishable as such; {to provide that no provisions of this Act shall be construed to repeal or limit certain laws or ordinances; to provide a severability clause;} to repeal conflicting laws; and for other purposes.

Be it enacted by the General Assembly of Georgia:

Section 1. No person shall, for the purpose of causing a condition of intoxication, stupefaction, euphoria, excitement, exhilaration, or dulling of the senses or nervous system, intentionally smell or inhale the fumes from any model glue as hereinafter defined; provided, that this section shall not apply to the inhalation of any anesthesia for medical or dental purposes.

Section 2. The term, "model glue," as used in this Act, shall mean any glue, cement, solvent or chemical substitute containing one or more of the following chemicals: Acetone, amyl chloride (iso- and tertiary), benzene, carbon disulfide, carbon tetrachloride, chloroform, ether, ethyl acetate, ethyl alcohol, ethylene dichloride, isopropyl acetate, isopropyl alcohol, isopropyl ether, methyl acetate, methyl alcohol, propylene dichloride, propylene oxide, trichloroethylene, amyl acetate, amyl alcohol, butyl acetate, butyl alcohol, butyl ether, diethylcarbonate, diethylene oxide (Dioxan), dipropyl ketone, ethyl butyrate, ethylene glycol monoethyl ether (Cellosolve), ethylene glycol monomethyl ether acetate (Methyl Cellosolve Acetate), isobutyl alcohol, methyl amyl acetate, methyl amyl alcohol, methyl isobutyl ketone, toluene.

Section 3. No persons shall intentionally possess, buy, sell, transfer possession or receive possession of any model glue for the purpose of violating or aiding another person to violate any provision of this Act.

Section 4. No person shall sell or transfer possession of any model glue to another person under eighteen (18) [twenty-one (21)] years of age nor shall any person under eighteen (18) [twenty-one (21)] years of age possess or buy any model glue unless the purchase is for model building or other lawful use and the person under eighteen (18) [twenty-one (21)] years of age has in his possession and exhibits to the seller or transferor the written consent of his parent or legal guardian to make said purchase or take possession of said model glue. {Provided, any minor who shall transfer possession of model glue to another minor for model building or other lawful purpose shall not be held criminally liable for failing to require exhibition of the written consent of the transferee-minor's parents or for failing to keep same available for inspection by law enforcement officials.}

Section 5. The person making a sale or transfer of possession of model glue to a person under eighteen (18) [twenty-one (21)] years of age must require such purchaser to exhibit the written consent of his parent or guardian and the name and address of the consenting parent or guardian. All data required by this section shall be kept available by the seller for inspection by law enforcement officials for a period of six months.

Section 6. Any person who shall violate any of the provisions of this Act shall, upon conviction thereof, be punished as for a misdemeanor. Each violation of any of the provisions of this Act shall be deemed to be a separate and distinct offense.

{Section 7. No provisions in this Act shall be construed to repeal or limit existing laws or ordinances of the governing authority of any county or municipality regulating, restricting or prohibiting the sale of model glue to any person under the age of eighteen, nor shall this Act restrict the governing authority of any county or municipality from enacting ordinances or regulations governing the regulation of model glue not inconsistent with this Act.

Section 8. In the event any section, subsection, sentence, clause or phrase of this Act shall be declared or adjudged invalid or unconstitutional, such adjudication shall in no manner affect the other Sections, Subsections, sentences, clauses, or phrases of this Act, which shall remain of full force and effect, as if the section, subsection, sentence, clause or phrase so declared or adjudged invalid or unconstitutional were not originally a part hereof. The General Assembly hereby declares that it would have passed the remaining parts of this Act if it had known that such part or parts hereof would be declared or adjudged invalid or unconstitutional.}

Section 9 [7]. All laws and parts of laws in conflict with this Act are hereby repealed.

Approved April 9, 1968.

## ALABAMA'S PROPOSED GLUE SNIFFING LEGISLATION

*A BILL*

*TO BE ENTITLED*

*AN ACT*

Relating to model glue; prohibiting inhalation and providing a penalty.

BE IT ENACTED BY THE LEGISLATURE OF ALABAMA:

Section 1. (A) As used in this act, the term "model glue" shall not mean any of the below set forth ingredients or types when used industrially and in the usual course of business, but shall mean any glue or cement of the type commonly used in the building of model airplanes, boats and automobiles and which contain one or more of the following volatile solvents:

(1). Acetone
(2). Amylacetate
(3). Benzol or benzene
(4). Butyl acetate
(5). Butyl alcohol
(6). Carbon tetrachloride
(7). Chloroform
(8). Cyclohexanone
(9). Ethanol or ethyl alcohol
(10). Ethyl acetate
(11). Hexane
(12). Isopropanol or isopropyl alcohol
(13). Isopropyl acetate
(14). Methyl "Cellosolve" acetate
(15). Methyl ethyl ketone
(16). Methyl isobutyl ketone
(17). Toluol or toluene
(18). Trichloroethylene
(19). Tricresyl phosphate
(20). Xylol or sylene

or any other solvent, material, substance, chemical or combination thereof, having the property of releasing toxic vapors.

*(B) It shall be unlawful for any person to intentionally smell or inhale the fumes of any type of "model glue" as defined in this act, or to induce any other person to do so, for the purpose of causing a condition of, or inducing symptoms of intoxication, elation, euphoria, dizziness, excitement, irrational behavior, exhilaration, paralysis, stupefaction or dulling of the senses of nervous system, or for the purpose of, in any manner, changing, distorting or disturbing the audio, visual or mental processes; provided, however, that this section shall not apply to the inhalation of any anesthesia for medical or dental purposes.*

*(C) It shall be unlawful for any person to intentionally possess any of "model glue" for the purpose of using the same in the manner prohibited by sub-section (b) of this section.*

*(D) It shall be unlawful for any person to intentionally possess, buy, sell or otherwise transfer any type of "model glue" for the purpose of inducing or aiding any other person to violate the provisions of sub-section (b) of this section.*

Section 2. *Any person who violates any of the provisions of this act shall upon conviction be guilty of a misdemeanor, punishable by a fine of not more than $500.00, or by imprisonment in the county jail for a term of not more than six months, or by both such fine and imprisonment in the discretion of the court.*

Section 3. *This act shall become effective immediately upon its passage and approval by the Governor, or upon its otherwise becoming law.*

# Appendix Three

## Profile of Glue Sniffers Across the Nation
### By Lester G. Thomas and the Denver Glue-Sniffing Project, 1967

*Included in the published proceedings of the 1967 Denver glue summit were the results of a national study commissioned by the Denver juvenile court, an attempt to compare the findings of Denver with the findings of other municipalities across the country.*

In order to determine whether Denver Court-acquainted glue-sniffers are essentially similar to their counterparts in other parts of the country, a nation-wide survey was conducted. A Questionnaire was sent to the Court responsible for juvenile delinquency cases in each jurisdiction in the nation with a total population of one hundred thousand or more. A total of 157 questionnaires were mailed and the gratifying total of 83 responses were received. This represents a 53% rate of response.

In terms of obvious characteristics, court-acquainted glue-sniffers appear to be similar throughout the country. The following represents responses to the nation-wide questionnaire (items in which Denver Court-acquainted glue-sniffers are different than national results, notes explaining the differences are included).

### Age

| | |
|---|---|
| 10 years or younger | 2% |
| 11 years | 8% |
| 12 years | 10% |
| 13 years | 14% |
| 14 years | 18% |
| 15 years | 27% |
| 16 years | 12% |
| 17 years | 8% |
| 18 years | 1% |
| | Total – 100% |

### Sex
| | |
|---|---|
| Male | 90% |
| Female | 10% |
| | Total – 100% |

### Race
| | |
|---|---|
| White | 66% |
| White with Spanish or Mexican surname | 29% |
| Negro | 4% |
| Other | 1% |
| | Total – 100% |

Note: In Denver, more than 90% of all known habitual glue-sniffers are white with Spanish surnames. Analysis of nation-wide results indicate that in the Rocky Mountain and Southwest regions, where a sizeable proportion of the population is Spanish or Mexican surnamed, the majority of known glue-sniffing children are of this ethnic background.

### Family income during past 12 months
| | |
|---|---|
| $0.00 to $999.00 | 5% |
| $1,000 to $2,999 | 26% |
| $3,000 to $4,999 | 53% |
| $5,000 to $6,999 | 15% |
| $7,000 to $9,999 | 1% |
| | Total – 100% |

Note: In Denver, approximately 40% of all Court-acquainted glue-sniffers come from families with annual incomes in the $1,000 to $2,999 category. Many of these families are welfare recipients or are dependent upon a marginally-employed father.

### Marital status of natural parents
| | |
|---|---|
| Married | 48% |
| Separated or divorced | 44% |
| Never married | 5% |
| Other | 3% |
| | Total – 100% |

## Percentage of glue-sniffing youngsters also involved in other delinquency: 80%

### School attendance of glue-sniffers is:
| | |
|---|---|
| Better than other delinquents | 13% |
| About the same as other delinquents | 53% |
| Not as good as other delinquents | 34% |
| | Total – 100% |

| | |
|---|---|
| Better than among other school children | 0% |
| About the same as other school children | 13% |
| About as good as other school children | 87% |
| | Total – 100% |

### Dropout rate among glue-sniffers is:
| | |
|---|---|
| Higher than among other delinquents | 14% |
| About the same as other delinquents | 68% |
| Lower than other delinquents | 18% |
| | Total – 100% |

| | |
|---|---|
| Higher than other school children | 59% |
| About the same as other school children | 33% |
| Not as high as other school children | 8% |
| | Total – 100% |

### Deportment grades of glue-sniffers are:
| | |
|---|---|
| Higher than other delinquents | 13% |
| About the same as other delinquents | 47% |
| Lower than other delinquents | 40% |
| | Total – 100% |

| | |
|---|---|
| Higher than other school children | 3% |
| About the same as other school children | 13% |
| Not as high as other school children | 84% |
| | Total – 100% |

## Relative achievement (grades) of glue-sniffers is:

| | |
|---|---|
| Higher than other delinquents | 3% |
| About the same as other delinquents | 63% |
| Lower than other delinquents | 34% |
| | Total – 100% |

| | |
|---|---|
| Higher than other school children | 0% |
| About the same as other school children | 9% |
| Not as high as other school children | 91% |
| | Total – 100% |

## Physical appearance

Glue-sniffing children:

| | |
|---|---|
| Have a more mature physical appearance than most delinquents of their age | 10% |
| Have about the same physical appearance as most delinquents of their age | 55% |
| Have a less mature physical appearance than most delinquents of their age | 35% |
| | Total – 100% |

| | |
|---|---|
| Have a more mature physical appearance than other school children of their age | 6% |
| Have about the same physical appearance as other school children of their age | 47% |
| Have a less mature physical appearance than other school children of their age | 47% |
| | Total – 100% |

Note: In Denver, virtually all Court-acquainted glue-sniffing children appear to be considerably less physically mature than other children of their age, delinquents or non-delinquents.

# Notes

## Notes to Introduction

1. Homer Bigart, "Slain Scout Aide Honored at Mass," *New York Times*, January 31, 1964, 32.

2. Sándor Radó, "The Psychoanalysis of Pharmacothymia (Drug Addiction)," *Psychoanalytic Quarterly* 2, no. 1 (1933): 8; Edward M. Brecher and the Editors of *Consumer Reports Magazine, The Consumers Union Report: Licit and Illicit Drugs*, http://www.druglibrary.org/Schaffer/LIBRARY/studies/cu/cumenu.htm, pt. 6, Inhalants, Solvents and Glue-sniffing (hereafter Brecher et al. *Consumers Union Report*), ch. 43, "The Historical Antecedents of Glue-Sniffing"; Charles William Sharp and Mary Lee Brehm, *Review of Inhalants: Euphoria to Dysfunction*, NIDA Research Monograph 15 (Department of Health, Education, and Welfare, October 1977), 2.

3. Brecher et al., *Consumers Union Report*, ch. 44, "How to Launch a Nationwide Drug Menace"; Sharp and Brehm, *Review of Inhalants*, 3–4.

4. Stanley Cohen, *Folk Devils and Moral Panics: The Creation of the Mods and Rockers* (1972; New York: St. Martin's Press, 1980); Kenneth Thompson, *Moral Panics* (New York: Routledge, 1998), 31–42; Erich Goode and Nachman Ben-Yehuda, "Moral Panics: Culture, Politics, and Social Construction," *Annual Review of Sociology* 20 (1994): 150, 154–56. Cohen's formulation included "folk devils," deviant stereotypes that became the focus of public concern and anger, a nonfactor in the glue-sniffing epidemic because the locus of concern were the victims of an intoxicant, the dealers were selling products that were intended to be moral, and the producers were making efforts to make their product less dangerous. For comparative history on Mods and Rockers, Efi Avdela described a similar phenomenon in Greece in the 1950s and early 1960s in Efi Avdela, "'Corrupting and Uncontrollable Activities': Moral Panic about Youth in Post–Civil War Greece," *Journal of Contemporary History* 43 (January 2008): 25–44.

5. Goode and Ben-Yehuda, "Moral Panics," 150–52. Such concern can be expressed through social movements, media stories, voting, or simple attitudes. Sociologists attempt to measure such concern through organized actions, amount of media coverage, or the availability of new laws.

6. Goode and Ben-Yehuda, "Moral Panics," 158. Moral panics also tend to include an increased level of hostility toward those affected, though this element is absent from the glue-sniffing epidemic. For an elaboration on Goode and Ben-Yehuda's arguments, see their book-length treatment of the subject in Erich Goode and Nachman Ben-Yehuda, *Moral Panics: The Social Construction of Deviance* (Oxford: Blackwell, 1994). Nachman Ben-Yehuda also effectively elaborates on some of this ground in his *Politics and Morality of Deviance: Moral Panics, Drug Abuse, Deviant Science, and Reversed Stigmatization* (Albany: SUNY Press, 1990); and Nachman Ben-Yehuda, "The Sociology of Moral Panics: Toward a New Synthesis," *Sociological Quarterly* 27 (Winter 1986): 495–513.

7. Brecher et al., *Consumers Union Report*, ch. 44; "L.I. Youths Inhale Glue in Model Kits for Narcotic Effect," *New York Times*, October 6, 1961, 37; "City Investigating 'Kicks' from Glue," *New York Times*, September 26, 1962, 24; "City Plans Drive on Glue-Sniffing," *New York Times*, April 25, 1963, 35; "Youth Killed in Plunge," *New York Times*, December 5, 1963, 52; "City Acts to Halt Sniffing of Glue," *New York Times*, December 18, 1963, 43. For more on Utah, see Donald E. Houseworth, "A Study of Retreatism in Glue-Sniffing and Non-Glue-Sniffing Delinquents in Utah" (PhD diss., Brigham Young University, 1968).

8. "Fads: The New Kick," *Time*, February 16, 1962, 55. See also "Slain Scout Aide Honored at Mass," *New York Times*, January 31, 1964, 32; "Store Owner Accused of Selling Glue to Boy Killed in Plunge," *New York Times*, March 3, 1964, 26; "Boy, Thirteen, Sniffing Glue, Falls into Gowanus and Drowns," *New York Times*, August 26, 1964, 16; "Curb on Glue Sales Urged," *New York Times*, August 30, 1964, 94; "Sniffing of Fluid Is Fatal to Boy; Four Others Made Ill," *New York Times*, June 19, 1965, 25; "Youth Dies after Sniffing Cleaning Fluid at Party," *New York Times*, January 9, 1965, 52; "Lacquer-Thinner Sniffing Blamed in Five Deaths in Japan," *New York Times*, June 15, 1967, 4.

9. Howard S. Becker, *Outsiders: Studies in the Sociology of Deviance* (New York: Free Press, 1963), 135–47; Kathleen Auerhahn, "The Split Labor Market and the Origins of Antidrug Legislation in the United States," *Law and Social Inquiry* 24 (Spring 1999): 432–36; Goode and Ben-Yehuda, "Moral Panics," 153–54. For more on moral entrepreneurs, motivations for action in panics, and the sociological theories that base understandings of those motivations, see Philip Jenkins, *Intimate Enemies: Moral Panics in Contemporary Great Britain* (New York: Aldine De Gruyter, 1992), 2–8; Sean P. Hier, "Tightening the Focus: Moral Panic, Moral Regulation and Liberal Government," *British Journal of Sociology* 62, no. 3 (2011): 523–41.

10. Goode and Ben-Yehuda, "Moral Panics," 167. If the panic is the result of middle-status organizations like the media, it is almost impossible to make the argument that there is not an interest-seeker involved, because organizational entities like the media don't have a moral status. Under the grassroots model of moral panics, there's no need for an interest group to initiate the panic, but it can facilitate it. The media isn't going to make something up. It will respond to the latent concerns of society. So the media isn't creating the panic, it is the grapevine upon which the grassroots concern is developing. See also Goode and Ben-Yehuda, "Moral Panics," 159–61, 163–64, 166; Gilbert Herdt, *Moral Panics, Sex Panics: Fear and the Fight over Sexual Rights* (New York: New York University Press, 2009). For the relationship between deviance and the media, see Robert Reiner, "Media Made Criminality: The Representation of Crime in the Mass Media," in *The Oxford Handbook of Criminology*, ed. Robert Reiner, Mike Maguire, Rod Morgan (New York: Oxford University Press, 2002), 302–40; Chris Greer, ed., *Crime and Media: A Reader* (New York: Routledge, 2009).

11. For more on the generational divide, parenting, and the kinds of misunderstandings that create fear among mothers and fathers, see Jacqueline Scott, "Is It a Different World to When You Were Growing Up? Generational Effects on Social Representations and Child-Rearing Values," *British Journal of Sociology* 51 (June 2000): 355–76.

12. John Springhall, *Youth, Popular Culture, and Moral Panics: Penny Gaffs to Gangsta Rap, 1830–1996* (New York: St. Martin's, 1999), 156.

13. Both were necessarily created in one way or another by the American consumerist culture, and, as Christopher Lasch noted a decade after the epidemic ran its course: "The propaganda of consumption turns alienation itself into a commodity. It addresses itself to

the spiritual desolation of modern life and proposes consumption as the cure." Christopher Lasch, *The Culture of Narcissism: American Life in an Age of Diminishing Expectations* (New York: W. W. Norton, 1978), 73. See also Daniel Horowitz, *The Anxieties of Affluence: Critiques of American Consumer Culture, 1939–1979* (Amherst: University of Massachusetts Press, 2004), 129–202; Joseph A. Kotarba and Andrea Fontana, eds., *The Existential Self in Society* (Chicago: University of Chicago Press, 1984).

14. "Port Huron Statement of the Students for a Democratic Society, 1962," accessed August 17, 2013, http://coursesa.matrix.msu.edu/~hst306/documents/huron.html. Durkheim noted that poverty itself encouraged restraint, which by "compelling moderation, accustoms men to it." Wealth, however, "suggests the possibility of unlimited success," thus leaving people to assume a lack of limits, leading them to rebellious despondency. Emile Durkheim, *Suicide* (New York: Free Press, 1951), 254. For more on the anomie of affluence, see Mark Abrahamson, "Sudden Wealth, Gratification, and Attainment: Durkheim's Anomie of Influence Reconsidered," *American Sociological Review* 45 (February 1980): 49–57; William Simon and John H. Gagnon, "The Anomie of Affluence: A Post-Mertonian Conception," *American Journal of Sociology* 82 (September 1976): 356–78.

15. Kenneth Keniston, *The Uncommitted: Alienated Youth in American Society* (New York: Harcourt Brace, 1965), 281–308 (307).

16. *Airplane!*, directed by Jim Abrahams, David Zucker, and Jerry Zucker (Paramount Pictures, 1980).

17. Michelle Alexander, *The New Jim Crow: Mass Incarceration in the Age of Colorblindness* (New York: New Press, 2010). See also King County Bar Association, "Drugs and the Drug Laws: Historical and Cultural Contexts," *Report of the Legal Frameworks Group to the King County Bar Association Board of Trustees* (Seattle: King County Bar Association Drug Policy Project, 2005), 29–31; Goode and Ben-Yehuda, "Moral Panics," 168–69. For more on the War on Drugs and further citations, see Conclusion.

18. Goode and Ben-Yehuda, "Moral Panics," 169–70.

## Notes to Chapter 1

1. A. A. Brill, "The Sense of Smell in the Neuroses and Psychoses," *Psychoanalytic Quarterly* 1 (April 1932): 7.

2. Ibid., 8.

3. Ibid., 9. Brill marshaled the research of everyone from British psychologist Havelock Ellis to German ethnologist Theodor Koch-Grünberg to make his case, patiently documenting the anecdotal evidence of the reliance on smell in the lives of civilizations viewed as fundamentally underdeveloped. Travel writer Frederick O'Brien argued that the noses of the South Sea Islanders "were sources of sensuous enjoyment to them beyond my capability. They inhaled emanations from flowers too subtle to touch my olfactory nerves." Arctic and Antarctic explorer Vilhjalmur Stefansson noted that "while in eyesight, hearing, and every other natural faculty," an Eskimo was "about the same as the rest of us, he does seem to excel in the sense of smell." Frederick O'Brien, *Atolls of the Sun* (New York: Century Co., 1922), 330. Vilhjalmur Stefansson, *My Life with the Eskimo* (New York: Macmillan, 1919), 64. See also Theodor Koch-Grünberg, *From Roraima to the Orinoco* (New York: Cambridge University Press, 2009); Havelock Ellis, *Sexual Selection in Man* (New York: F. A. Davis, 1906), 65–76.

4. Ellis, *Sexual Selection*, 70; Richard von Krafft-Ebing, *Psychopathia Sexualis* (1886; Burbank, CA: Bloat Books, 1999); Connolly Norman, "Variations in Form of Mental Affections in Relation to the Classification of Insanity," *Dublin Journal of Medical Science* 83 (1887): 228–35; Eugen Bleuler, *Lehrbuch de Psychiatrie* (Berlin: Springer-Verlag, 1916), 48; and Brill, "Sense of Smell," 10–12.

5. Brill, "Sense of Smell," 13.

6. Sigmund Freud, "Bemerkungen über einen Fall von Zwangsneurose," *Gesammelte Schriften* 8 (1908): 350. Along with that work, translated as "Notes upon a Case of Obsessional Neurosis," Freud also discussed the link between smell and sexual development in Sigmund Freud, *Three Contributions to the Theory of Sex*, trans. A. A. Brill (New York: Nervous and Mental Disease Publishing Co., 1920).

7. Brill, "Sense of Smell," 15.

8. Ibid., 12–15; von Krafft-Ebing, *Psychopathia Sexualis*, 31. See Wilhelm Fliess, *Die Beziehungen zwischen Nase und weiblichen Geschlechtsorganen, In ihrer biologischen Bedeutung dargestellt* (Leipzig: Deuticke, 1897), which translates as "The relationship between the nose and female sex organs, its biological significance shown."

9. Brill, "Sense of Smell," 17–26.

10. Ibid., 25.

11. Ibid., 37.

12. Ibid., 30–31, 28, 41, 37. See also Somerset Maugham, *On a Chinese Screen* (New York: Doran and Co., 1922), 140; Oscar Wilde, *The Picture of Dorian Gray* (London: Ward, Lock, and Co., 1891); Leopold Bernard, *Les odeurs dans les romans de Zola* (Montpellier: C. Coulet, 1889).

13. Brill, "Sense of Smell," 37.

14. Ibid., 39.

15. Ibid.

16. Gare Hambridge, "Scents That Make Dollars," *World's Work* 62 (August 1931): 274; Brill, "Sense of Smell," 33; Sándor Ferenczi, *First Contributions to Psycho-Analysis*, trans. Ernest Jones (1916; London: Karnac Books, 1994).

17. John A. Nunn and Frank M. Martin, "Gasoline and Kerosene Poisoning in Children," *Journal of the American Medical Association* (hereafter *JAMA*) 103 (August 18, 1934): 472–73.

18. Ibid., 473–74.

19. Willard Machle, "Gasoline Intoxication," *JAMA* 117 (December 6, 1941): 1965; Becker, *Outsiders*, 135–47.

20. Radó, "Psychoanalysis of Pharmacothymia," 2.

21. Ibid., 1–3. It should come as no surprise that there was an "etiological importance of the erotogenic oral zone and a close relationship to homosexuality" for Radó, building as his work (and everyone's work) did off of that of Sigmund Freud, Freud's student Viktor Tausk, Heinz Hartmann, and other European psychological luminaries. There was, without fail, always a sexual component to every form of narcotic-induced neurosis.

22. Ibid., 4.

23. Ibid., 4–5, 18–19. And that self-destruction almost always included a "weakening of genital masculinity" and, ultimately, homosexuality, the common fear of all psychoanalysts dealing with problems of addiction in the early twentieth century.

24. Ibid., 6.

25. Ibid., 7.

26. Ibid., 9, 11. Even when children were involved, sexuality was an issue. Such drug-induced pleasure "initiates an artificial sexual organization which is autoerotic and modeled on infantile masturbation." Sex becomes a regressive object of fantasy "to the emotional attachments of childhood, that is to say, to the Oedipus complex."

27. Ibid., 16–17.

28. Machle, "Gasoline Intoxication," 1968.

29. Ibid., 1965–68. Of course, the potentially toxic makeup of products that people used daily for other needs was ever a concern. In the 1950s, the *JAMA* published a report from Community Memorial Hospital in South Hill, Virginia, on the dangers of nutmeg poisoning. The common flavoring agent was formed from 5–15 percent volatile oils, including myristicin (5-allyl-1-methoxy-2,3-methylenedioxybenzene), dextrocamphene, dextropinene, and dipentene, all of which could prove dangerous when consumed in large quantities. See Robert C. Green Jr. "Nutmeg Poisoning," *JAMA* 171 (November 7, 1959): 1342–44.

30. Rex H. Wilson, "Toluene Poisoning," *JAMA* 123 (December 25, 1943): 1106.

31. Leonard Greenburg, May R. Mayers, Harry Heimann, and Samuel Moskowitz, "The Effects of Exposure to Toluene in Industry," *JAMA* 118 (February 12, 1942): 573; Wilson, "Toluene Poisoning," 1106.

32. Greenburg et al., "Toluene in Industry," 573–78; Wolfgang F. von Oettingen, Paul A. Neal, and Dennis D. Donahue, "The Toxicity and Potential Dangers of Toluene," *JAMA* 118 (February 12, 1942): 579–84. The munitions study measured exposure to toluene at concentrations of between 50 and 800 parts per million.

33. Wilson, "Toluene Poisoning," 1106–8. The study of the industrial plant measured exposure between 50 and 1,500 parts per million.

34. H. B. Lockhead and H. P. Close, "Ethylene, Dichloride Plastic Cement: A Case of Fatal Poisoning," *JAMA* 146 (August 4, 1951): 1323. That is not to say that accidental exposure concerns just disappeared. Ethylene dichloride ($CH_2Cl$-$CH_2Cl$) was a chlorinated hydrocarbon used in plastic cement, thereby making it prominent in industrial use, occupational therapy, and "in homes where work with plastics is a hobby." Though inhalation of the substance had not caused any reported cases of poisoning, ingestion certainly had. In the majority of cases, people drank the substance, mistaking it for liquor. It is significant that worries about plastic cement were available as early as 1951, but again the focus was specifically on accidental ingestion, not intentional inhalation.

35. Orris W. Clinger and Nelson A. Johnson, "Purposeful Inhalation of Gasoline Vapors," *Psychiatric Quarterly* 25 (October 1951): 557. See also J. M. Schneck, "Chloroform Habituation: With a Case Report of Its Occurrence in a Case of Schizophrenia," *Bulletin of the Menninger Clinic* 9 (January 1945): 12–17.

36. Clinger and Johnson, "Gasoline Vapors," 558; Robert L. Faucett and Reynold A. Jensen, "Psychologic Aspects of Pediatrics: Addiction to the Inhalation of Gasoline Fumes in a Child," *Journal of Pediatrics* 41 (September 1952): 364. Such was not the first recorded case. In 1900, there was a "14-year-old feeble-minded girl" who became problematically irritable after inhaling benzene fumes. At the time, however, the case was deemed to be an isolated incident based on her previous mental status.

37. Clinger and Johnson, "Gasoline Vapors," 560.

38. Ibid., 558, 560–62.

39. Ibid., 561–63.

40. Ibid., 561–62.

41. Ibid., 562–63.
42. Ibid., 562–64.
43. Ibid., 564, 567.
44. Faucett and Jensen, "Aspects of Pediatrics," 366.
45. Ibid., 364–68. The thrust of Faucett and Jensen's work in this case was to convince people of the validity of using sodium amytal to open the minds of children to repressed memories or hallucinations.
46. Donald Gerard and Conan Kornetsky, "Adolescent Opiate Addiction," *Psychiatric Quarterly* 29 (July 1955): 457.
47. Ibid., 457–60. "Residence in unsanitary, restricted, crowded, over-priced housing; prejudicial limitations to occupational advancement, and rejection by white teachers working under pressure in understaffed, over-crowded schools, are some of the gross social experiences which intensify the problems of adolescent adjustment for Negro and Puerto Rican youth." With that understanding in tow, Gerard and Kornetsky emphasized only black and Puerto Rican subjects in their study.
48. Ibid., 484.
49. Ibid., 460, 464, 468–74, 484. For more on the theory of childhood and child development and how thinking about children and their status in society has developed over time, see Stuart Aitken, *Geographies of Young People: The Morally Contested Spaces of Identity* (New York: Routledge, 2001), 27–61 in particular.
50. Isidor Chein and Eva Rosenfeld, "Juvenile Narcotics Use," *Law and Contemporary Problems* 22 (Winter 1957): 65.
51. Ibid., 52–68.
52. Paul Friedman, "Some Observations on the Sense of Smell," *Psychoanalytic Quarterly* 28 (July 1959): 307–29.

## Notes to Chapter 2

1. Brecher et al., *Consumers Union Report*, ch. 44; Betty J. Fluke and Lillian R. Donato, "Some Glues Are Dangerous," *Empire Magazine*, supplement to *Denver Post*, August 2, 1959, 24. For more on Masterson, see Robert K. DeArment, *Bat Masterson: The Man and the Legend* (Norman: University of Oklahoma Press, 1989). For more on Wilson, see John Milton Cooper, *Woodrow Wilson: A Biography* (New York: Alfred A. Knopf, 2009).
2. US Bureau of the Census, *1970 Census of Population*, vol. 1, *Characteristics of the Population*, part A, *Number of Inhabitants*, section 1, United States, Alabama-Mississippi (Washington, DC: US Government Printing Office [hereafter USGPO], 1972), 7–13, 7–22, 7–23; US Bureau of the Census, *1970 Census of Population, Subject Reports: Final Report PC(2)-9B: Low-Income Areas in Large Cities* (Washington, DC: USGPO, 1973), 175, 177–78.
3. US Bureau of the Census, *1970 Census of Population*, 1:1–85, 4–15.
4. Brecher et al., *Consumers Union Report*, ch. 44; Fluke and Donato, "Some Glues Are Dangerous," 24. For more on Stanford's CIA experiment, see "Project MKULTRA, The CIA's Program of Research into Behavioral Modification. Joint Hearing before the Select Committee on Intelligence and the Subcommittee on Health and Scientific Research of the Committee on Human Resources," US Senate, Ninety-Fifth Congress, First Session, August 8, 1977 (Washington: USGPO, 1977).
5. Fluke and Donato, "Some Glues Are Dangerous," 24.

6. "Could Be Fatal: Plane Glue Gives Kids a Kick," *Denver Post*, June 12, 1960, Denver Public Library, TN#94979.

7. For more on the role of the media and agency in the generation of moral panics, see Arnold Hunt, "'Moral Panic' and Moral Language in the Media," *British Journal of Sociology* 48 (December 1997): 629–48.

8. Joseph F. Rorke, "Plastic Cement Fumes," *JAMA* 173 (July 16, 1960): 1277; Maurice Pruitt, "Bizarre Intoxications," *JAMA* 171 (December 26, 1959): 2355. The *JAMA* had certainly dealt with inhalation problems before. As the glue-sniffing epidemic first developed in Denver, the journal published a letter to the editor dealing with the continuing problem of gasoline sniffing as well as the use of nutmeg for intoxication. The notion that people, even children, would find available household products and use them to get high was far from new. See chapter 1.

9. William Shakespeare, "The Winter's Tale," act 3, scene 3, accessed August 4, 2011, http://www.classicreader.com/book/164/9/; Fritz Redl, "The Psychology of Gang Formation and the Treatment of Juvenile Delinquents," in *Psychoanalytic Study of the Child* (New York: International Universities Press, 1945), 1:367. Redl studied under August Aichhorn, one of the principal founders of child psychology and the author of *Wayward Youth* (New York: Viking Press, 1935), among others.

10. Melitta Schmideberg, "The Psychoanalysis of Delinquents," *American Journal of Orthopsychiatry* 23 (January 1953): 13.

11. Ibid., 13–21; also Alfred E. Coodley, "Current Aspects of Delinquency and Addictions," *Archives of General Psychiatry* 4 (June 1961): 632–34.

12. Coodley, "Delinquency and Addictions," 634.

13. Ibid.

14. Ibid., 635.

15. Ibid., 632.

16. Ibid., 633; also Ednita P. Bernabeu, "Underlying Ego Mechanisms in Delinquency," *Psychoanalytic Quarterly* 27 (July 1958): 383–96.

17. Coodley, "Delinquency and Addictions," 636–39. This problematic vagueness in defining addiction was pointed out more popularly by Joel Fort in *The Pleasure Seekers: The Drug Crisis, Youth, and Society* (Indianapolis: Bobbs-Merrill, 1969), 8–9.

18. James J. Lawton and Carl P. Malmquist, "Gasoline Addiction in Children," *Psychiatric Quarterly* 35 (July 1961): 555–56.

19. Ibid., 557.

20. Ibid., 558.

21. Ibid., 558. The Minnesota study fretted about the lack of literature existing on intentional gasoline inhalation, but although the literature was spotty it did exist. See Clinger and Johnson, "Gasoline Vapors"; Faucett and Jensen, "Aspects of Pediatrics," 364–367; C. J. Nitsche and J. F. Robinson, "A Case of Gasoline Addiction," *American Journal of Orthopsychiatry* 24 (April 1959): 417–19; Robert V. Edwards, "A Case Report of Gasoline Sniffing," *American Journal of Psychiatry* 117 (December 1960): 555–57.

22. Lawton and Malmquist, "Gasoline Addiction in Children," 559–60.

23. Bill Miller, "Seven Denver Youths Arrested for Deadly 'Glue-Sniffing,'" *Denver Post*, October 23, 1961, Denver Public Library, TN#94980.

24. Ibid.

25. Lenny Bruce, "Airplane Glue," *American*, Fantasy Records, 7011, F-2081, 1961. The Avro Lancaster airplane was a British heavy bombing plane used during World War II.

George Macready was an American actor known for playing villains in Hollywood films. For more on Bruce, his life and comedy, see Lenny Bruce, *How to Talk Dirty and Influence People* (New York: Playboy Publishing, 1963).

26. For more on Bruce's obscenity trials, see Ronald K. L. Collins and David M. Skover, *The Trials of Lenny Bruce: The Fall and Rise of an American Icon* (New York: Sourcebooks, 2002).

27. Jim Ritchie, "Law Asked on Sniff 'Narcotics,'" *Denver Post*, December 8, 1961.

28. "Public Law 86-613, July 12, 1960," *United States Statutes at Large, 1960*, vol. 74 (Washington, DC: USGPO, 1961), 372.

29. Ibid., 372-81. The law was originally known as the Federal Hazardous Substances Labeling Act, but the word "labeling" was later removed from the title.

30. Al Arnold, "Glue-Sniffing Rage Grows," *Denver Post*, April 29, 1962, 1.

31. Ibid. See also John Kokish, "Denver Youth Crime Up 18 Pct. in 1961," *Denver Post*, April 13, 1962, 1, 43.

32. William M. Easson, "Gasoline Addiction in Children," *Pediatrics* 29 (February 1962): 253.

33. Ibid.

34. Ibid., 250-54.

35. Helen H. Glaser and Oliver N. Massengale, "Glue-Sniffing in Children: Deliberate Inhalation of Vaporized Plastic Cements," *JAMA* 181 (July 28, 1962): 300.

36. Ibid.

37. Glaser and Massengale, "Glue-Sniffing in Children," 301-2. See also Marion N. Gleason, Robert E. Gosselin, and Harold Carpenter Hodge, *Clinical Toxicology of Commercial Products* (Baltimore: Williams and Wilkins, 1957); Robert H. Dreisbach, ed., *Handbook of Poisoning: Diagnosis and Treatment* (Los Altos, CA: Lange Medical Publications, 1961); M. A. Wolf, V. K. Row, D. D. McCollister, R. L. Hollingsworth, and F. Oyen, "Toxicological Studies of Certain Alkylated Benzenes and Benzene," *Archives of Industrial Health* 14 (October 1956): 387-98.

38. Glaser and Massengale, "Glue-Sniffing in Children," 303.

39. Ibid., 301.

40. Ibid.

41. Ibid.; "Glue-Sniffing," *JAMA* 181 (July 28, 1962): 333.

42. Eric C. Schneider, *Smack: Heroin and the American City* (Philadelphia: University of Pennsylvania Press, 2008), 116-24.

43. Timothy A. Hickman, *The Secret Leprosy of Modern Days: Narcotic Addiction and Cultural Crisis in the United States, 1870-1920* (Amherst: University of Massachusetts Press, 2007), 68; David F. Musto, *The American Disease: Origins of Narcotic Control* (1973; New York: Oxford University Press, 1987), 11; David T. Courtwright and Timothy A. Hickman, "Modernity and Anti-modernity: Drug Policy and Political Culture in the United States and Europe in the Nineteenth and Twentieth Centuries," in *Drugs and Culture: Knowledge, Consumption, and Policy*, ed. Geoffrey Hunt, Maitena Milhet, and Henry Bergeron (Burlington, VT: Ashgate, 2011), 215.

44. Hickman, *Secret Leprosy*, 72-80; Musto, *American Disease*, 11; Courtwright and Hickman, "Modernity and Anti-modernity," 218-19. This kind of assumption would continue about cocaine throughout the twentieth century, particularly after its production in crack form.

45. Fort, *Pleasure Seekers*, 13.

46. "Glue Sniffing: An Adolescent Craze That Is Not Amusing," *Consumer Reports* 28 (January 1963): 40.

47. Ibid.

## Notes to Chapter 3

1. Michael J. Whelton, "Glue Sniffing," *British Medical Journal* 2 (November 24, 1962): 1404. Whelton was referring to the Single Convention on Narcotic Drugs, an international drug treaty signed in 1961 that placed restrictions on international opiate sales without prescriptions. The United States actually began the regulation of opiates such as codeine in the 1914 Harrison Narcotics Act. Regardless, the availability of such products (and others previously mentioned like turpentine and gasoline) make Whelton's explanation for the onset of the glue-sniffing epidemic inadequate. See also "Single Convention on Narcotic Drugs, 1961," United Nations, accessed July 26, 2013, http://www.incb.org/documents/Narcotic-Drugs/1961-Convention/convention_1961_en.pdf; "Harrison Narcotics Act (1914)," *Public Acts of the Sixty-Third Congress of the United States*, 38 United States Statutes at Large 785, chapter 1, December 17, 1914; Saran Ghatak, "The Opium Wars: The Biopolitics of Narcotic Control in the United States, 1914–1935," *Critical Criminology* 18 (March 2010): 41–56; Kurt Hohenstein, "Just What the Doctor Ordered: The Harrison Anti-narcotic Act, the Supreme Court, and the Federal Regulation of Medical Practice, 1915–1919," *Journal of Supreme Court History* 26, no. 3 (2001): 231–56.

2. "L.I. Youths Inhale Glue in Model Kits for Narcotic Effect," *New York Times*, October 6, 1961, 37.

3. Ibid. See also Brecher et al., *Consumers Union Report*, ch. 44.

4. Leslie H. Whitten, "'Glue Sniffing' among Pupils Spreads Alarm," *Washington Post*, January 21, 1962, B1.

5. Ibid.

6. Ibid.

7. "History of the Testor Corporation," Reference for Business: Company History Index, accessed June 14, 2011, http://www.referenceforbusiness.com/history2/77/The-Testor-Corporation.html; "55 Years at Testor," *Rockford Register Star*, May 12, 1994, 1.

8. Whitten, "'Glue Sniffing' among Pupils Spreads Alarm," *Washington Post*, January 21, 1962, B1; also "History of the Testor Corporation."

9. "Fads: The New Kick," *Time*, February 16, 1962, 55.

10. Brecher et al., *Consumers Union Report*, ch. 44; "Fads," 55. For more on the state of glue sniffers in Utah, see Housworth, "A Study of Retreatism."

11. "Fads," 55.

12. Brecher et al., *Consumers Union Report*, ch. 44; Philip E. Norton, "Perilous Teen-Age Fad, Glue-Sniffing, Alarms Hobby Kit Industry," *Wall Street Journal*, December 7, 1962, 1.

13. Dorothy Gilliam, "Glue-Sniffing Causes Concern," *Washington Post*, November 11, 1962, L4.

14. "The New Addicts," *Newsweek*, August 13, 1962, 42.

15. Ibid. *Newsweek* would wring its hands over a similar fad the following year, when it reported that nursery owners were selling morning glory seeds at up to fifty times their average. The reason was that the seeds, when taken in bulk, could produce a hallucinogenic effect

similar to LSD. Morning glory varieties Heavenly Blue, Flying Saucers, and Pearly Gates all seemed to produce the high. The problem was most severe in San Francisco, Boston, and New York, leading the US Food and Drug Administration to begin its own investigation of the seemingly innocent flower. See "Going to Seed," *Newsweek*, July 22, 1963, 59.

16. Julius Merry and Nicholas Zachariadis, "Addiction to Glue Sniffing," *British Medical Journal* 2 (December 1, 1962): 1448.

17. Lawrence O'Kane, "City Plans Drive on Glue-Sniffing," *New York Times*, April 25, 1963, 35.

18. Ibid., 35. See also Brecher et al., *Consumers Union Report*, ch. 44.

19. Both the Anaheim and Maryland laws were published in *Bulletin (National Clearinghouse for Poison Control Centers)* (July–August 1964): 1. See also Brecher et al., *Consumers Union Report*, ch. 44. For more on the machinations that created glue-sniffing legislation and the various legal remedies created by states to solve such problems, see chapter 8.

20. Joseph F. Spillane, *Cocaine: From Medical Marvel to Modern Menace in the United States, 1884–1920* (Baltimore: Johns Hopkins University Press, 2000), 123–28.

21. Gilliam, "Glue-Sniffing Causes Concern," *Washington Post*, November 11, 1962, L4; "Four on 'Glue-Sniffing' Jag Steal Fifteen Cars," *Washington Post*, December 13, 1962, A3.

22. Paul A. Schuette, "Glue-Sniffing by Teen-Agers Faces City Ban," *Washington Post*, January 23, 1963, A3.

23. Theodore R. Van Dellen, "How to Keep Well," *Washington Post*, February 9, 1963, E17; "Radio Highlights for the Week," *Washington Post*, February 3, 1963, G5; "Police Assistance Is Sought Oftener," *Washington Post*, February 27, 1963, C1; "Hearing Will Consider 'Glue-Sniffing' Danger," *Washington Post*, June 7, 1963, B6.

24. Paul A. Schuette, "Proposed Curbs on Glue-Sniffing Bring Snorts from Hobby Lobby," *Washington Post*, July 20, 1963, D1. Joining Blyer in protest were Anis Amary, editor of *Toy and Hobby World* magazine, and representatives of the Petra Chemical Company.

25. "Navy Objects to Ban on Glue Sale," *Washington Post*, August 1, 1963, B2.

26. Alfred E. Lewis, "Juvenile Arrests Climb 13% in 1963," *Washington Post*, August 16, 1963, C1; "D.C. Vote Plan Called 'Superficial,'" *Washington Post*, November 24, 1963, B2; Paul A. Schuette, "Curb Proposed on Sale of Glue," *Washington Post*, December 18, 1963, C1.

27. "Glue-Sniffers in D.C. Face Jail and $300 Fine," *Washington Post*, February 19, 1964, C1.

28. "Glue Sniffing: An Adolescent Craze That Is Not Amusing," *Consumer Reports* 28 (January 1963): 40. See also Glaser and Massengale, "Glue-Sniffing in Children," 300–303.

29. Irving Spiegel, "Mayor Asks U.S. for Narcotics Aid," *New York Times*, June 4, 1963, 16.

30. "Youth Killed in Plunge," *Washington Post*, December 5, 1963, 52; Sydney H. Schanberg, "City Acts to Halt Sniffing of Glue," *Washington Post*, December 18, 1963, 43.

31. Schanberg, "City Acts to Halt Sniffing of Glue," *Washington Post*, December 18, 1963, 43; "Store Owner Accused of Selling Glue to Boy Killed in Plunge," *Washington Post*, March 3, 1964, 26.

## Notes to Chapter 4

1. For more, see McKenzie Andre, Kashef Ijaz, Jon D. Tillinghast, Valdis E. Krebs, Lois A. Diem, Beverly Metchock, Theresa Crisp, and Peter D. McElroy, "Transmission Network Analysis to Complement Routine Tuberculosis Contact Investigations," *American Journal of Public Health* 97 (March 2007): 470.

2. Gordon H. Barker and W. Thomas Adams, "Glue Sniffers," *Sociology and Social Research* 47 (April 1963): 298–99.

3. Ibid., 299.

4. Ibid., 303. That said, such were only hypotheses. Perhaps rural communities were less likely to send their children to reform schools. Perhaps rural communities provided less danger of being caught. Perhaps rural communities suppressed reports of glue sniffing so as not to draw negative attention to their populations or scare area parents. These possibilities were not explored in Barker and Adams's "Glue Sniffers."

5. Ibid, 304.

6. Ibid., 305.

7. Ibid., 307. Sokol submitted a report of the glue-sniffing problems in Los Angeles on August 14, 1962, to the Public Health Service of the US Department of Health, Education, and Welfare.

8. James W. Sterling, "A Comparative Examination of Two Modes of Intoxication: An Exploratory Study of Glue Sniffing," *Journal of Criminal Law, Criminology, and Political Science* 55 (March 1964): 94.

9. Ibid., 95–96.

10. Ibid., 96–98.

11. Barker and Adams, "Glue Sniffers," 307–8; Sterling, "Two Modes of Intoxication," 99.

12. Glaser and Massengale, "Glue-Sniffing in Children," 300–303; H. Jacobziner and H. W. Raybin, "Glue-Sniffing," *New York State Journal of Medicine* 62 (October 15, 1962): 3294; Wilson, "Toluene Poisoning"; Per Anderson and Birger R. Kaada, "The Electroencephalogram in Poisoning by Lacquer Thinner," *Acta Pharmacologica et Toxicologica* 9 (April 1953): 125; D. A. Grabski, "Toluene Sniffing Producing Cerebellar Degeneration," *American Journal of Psychiatry* 118 (November 1961): 461.

13. Josiah Dodds and Sebastiano Santostefano, "A Comparison of the Cognitive Functioning of Glue Sniffers and Non-sniffers," *Journal of Pediatrics* 64 (April 1964): 565–69. Similarly, a study from the Department of Medicine at General Hospital of Fresno County, California, came up with the same conclusions. They witnessed no significantly abnormal neurological findings, no hepatomegaly or splenomegaly, no renal failure, no bone marrow depletion. See Martin L. Barman and Donn B. Beedle, "Acute and Chronic Effects of Glue Sniffing," *California Medicine* 100 (January 1964): 19–22.

14. Dodds and Santostefano, "Comparison," 569.

15. Howard W. Pierson, "Glue Sniffing: A Hazardous Hobby," *Journal of School Health* 34 (May 1964): 252; Darleen Powars, "Aplastic Anemia Secondary to Glue Sniffing," *New England Journal of Medicine* 273 (September 23, 1965): 700–701. In addition, the National Clearinghouse for Poison Control Centers also demonstrated the prevalence of hypoplastic bone marrow and aplastic anemia. For the Poison Control warning, see Henry L. Verhulst and John J. Crotty, "Glue-Sniffing, II," *Bulletin (National Clearinghouse for Poison Control Centers)* (July–August 1964): 5.

16. Powars, "Aplastic Anemia," 701-2. See also S. J. Latta and L. Davies, "Effects on the Blood and Hemopoietic Organs of the Albino Rat of Repeated Administration of Benzene," *Archives of Pathology and Laboratory Medicine* 31 (1941): 55-67.

17. Jacob Sokol, "Glue Sniffing among Juveniles," *American Journal of Correction* 27 (November-December 1965): 18. See also Verhulst and Crotty, "Glue-Sniffing, II," 5.

18. Sokol, "Among Juveniles," 20. See also Powars, "Aplastic Anemias," 701-2.

19. Willie S. Ellison, "Portrait of a Glue Sniffer," *Crime and Delinquency* 11, no. 4 (1965): 394-97.

20. Council of Economic Advisers, *Economic Report of the President, Transmitted to the Congress February 1992* (Washington, DC: USGPO, 1992), 341. Even with the fluctuating unemployment, the United States remained an acquisitive culture, as consumers sought to fill an economic gap socially. See Christine Zumello, "The 'Everything Card' and Consumer Credit in the United States in the 1960s," *Business History Review* 85 (Autumn 2011): 551-75.

21. Scott Moody, ed., *Facts and Figures on Government Finance*, 33rd edition (Washington, DC: Tax Foundation, 1999), 48; Harold W. Stanley and Richard G. Niemi, *Vital Statistics on American Politics, 2011-2012* (Thousand Oaks, CA: CQ Press, 2011), 364; Bruce A. Chadwick and Tim B. Heaton, *Statistical Handbook on the American Family* (Phoenix: Oryx Press, 1999), 255; Robert Shaffer, "Public Employee Unionism: A Neglected Social Movement of the 1960s," *History Teacher* 44 (August 2011): 491-93. Such numbers also led to a significant increase in membership for public employee unions and a corresponding growth in the militancy of those unions. Of course, the history of class in the political economy of the United States is well-trodden ground. Perhaps the most influential document in that vein is Michael Harrington, *The Other America: Poverty in the United States* (1962; New York: Penguin, 1981). Other useful contextual works are Lucile Duberman, *Social Inequality: Class and Caste in America* (Philadelphia: J. B. Lippincott, 1976); Daniel W. Rosides, *The American Class System: An Introduction to Social Stratification* (Boston: Houghton Mifflin, 1976); also Peter d'A. Jones, *The Consumer Society: A History of American Capitalism* (New York: Penguin, 1965), 338-58.

22. US Bureau of the Census, *1970 Census of Population, Subject Reports: Final Report PC(2)-9A: Low-Income Population* (Washington, DC: USGPO, 1973), 154-56. See also Max Rose and Frank R. Baumgartner, "Framing the Poor: Media Coverage and US Poverty Policy, 1960-2008," *Policy Studies Journal* 41 (February 2013): 22-53; Andrew Gelman, Lane Kenworthy, and You-Sung Su, "Income Inequality and Partisan Voting in the United States," *Social Science Quarterly* 91 (December 2010): 1203-19. For more on the intersection of race and class, particularly as it relates to parents' protest movements (in this case welfare rights and Boston busing), see Georgina Denton, "'Neither Guns nor Bombs—Neither the State nor God—Will Stop Us from Fighting for Our Children': Motherhood and Protest in 1960s and 1970s America," *Sixties: A Journal of History, Politics, and Culture* 5 (December 2012): 205-28.

This racial bifurcation happened in other areas, too. The 1960s, for example, was the decade when advertisers, playing on themes of masculinity, steered white males to smoke king cigarettes and black males to smoke menthols. See Cameron White, John L. Oliffe, and Joan L. Bottorff, "From Promotion to Cessation: Masculinity, Race, and Style in the Consumption of Cigarettes, 1962-1972," *American Journal of Public Health* 103 (April 2013): 44-55.

23. Ellison, "Portrait," 397-98; Sokol, "Among Juveniles," 20; Powars, "Aplastic Anemia," 702.

24. Expert Committee on Drugs Liable to Produce Addiction, *Expert Committee on Drugs Liable to Produce Addiction: Second Report*, World Health Organization Technical Report Series No. 21 (Geneva: World Health Organization, 1950), 6–7; Alfred R. Lindesmith, "A Sociological Theory of Drug Addiction," *American Journal of Sociology* 43 (January 1938): 593–613; Alfred R. Lindesmith, *Addiction and Opiates* (1947; Chicago: Aldine Publishing, 1968).

25. Ellison, "Portrait," 397–98; Leland M. Corliss, "A Review of the Evidence on Glue-Sniffing—A Persistent Problem," *Journal of School Health* 35 (December 1965): 443–44; Edward Press, "Glue Sniffing," *Journal of Pediatrics* 63 (September 1963): 517; Alfred M. Freedman and Ethel A. Wilson, "Childhood and Adolescent Addictive Disorders," *Pediatrics* 34 (1964): 283–84.

26. Press, "Glue Sniffing," 517.

27. The principal study of Vietnam veterans' heroin use was conducted by Lee Robins under the leadership of Jerome H. Jaffe, leader of the Special Action Office of Drug Abuse Prevention in the Nixon administration. For more on this subject, see Jeremy Kuzmarov, *The Myth of the Addicted Army: Vietnam and the Modern War on Drugs* (Amherst: University of Massachusetts Press, 2009); Norman E. Zinberg, *Drug, Set, and Setting: The Basis for Controlled Intoxicant Use* (New Haven: Yale University Press, 1986); Norman E. Zinberg and John A. Robertson, *Drugs and the Public* (New York: Simon and Schuster, 1972), 58–86; Lee N. Robins, "Vietnam Veterans' Rapid Recovery from Heroin Addiction: A Fluke or Normal Expectation?" *Addiction* 88 (August 1993): 1041–54. See also Joint Committee of the American Bar Association and the American Medical Association on Narcotic Drugs, *Drug Addiction: Crime or Disease? Interim and Final Reports* (Bloomington: University of Indiana Press, 1961).

28. Nicolas Rasmussen, *On Speed: The Many Lives of Amphetamine* (New York: New York University Press, 2008), 197.

29. Ibid., 197–99. In their studies of crack users in the 1980s, Craig Reinarman and his colleagues noted that users often binged with long periods in between use. It was the kind of behavior that model airplane glue users had exhibited twenty years prior, and as Reinarman noted, "Binges are not addiction and do not by themselves cause addiction." Quote from Craig Reinarman, Dan Waldorf, Sheigla B. Murphy, and Harry G. Levine, "The Contingent Call of the Pipe: Bingeing and Addiction among Heavy Cocaine Smokers," in *Crack in America: Demon Drugs and Social Justice*, ed. Craig Reinarman and Harry G. Levine (Berkeley: University of California Press, 1997), 78.

30. Oliver Oldschool, "Ether Sniffers," *JAMA* 197 (July 11, 1966): 13.

31. William C. Ackerly and Guadalupe Gibson, "Lighter Fluid 'Sniffing,'" *American Journal of Psychiatry* 120 (May 1964): 1056.

32. Fort, *Pleasure Seekers*, 99–100; Oldschool, "Ether Sniffers," 13; Ackerly and Gibson, "Lighter Fluid 'Sniffing,'" 1056–61. For more on the white use of model airplane glue, see Rosario Anthony Caputo, *Hillbilly Glue Sniffing and Delinquency: The Social Structuring of Deviant Behavior among Southern White Migrants* (EdD diss., Columbia University Teachers College, 1988).

33. Frederick Glaser, "Inhalation Psychosis and Related States," *Archives of General Psychiatry* 14 (March 1966): 316.

34. John Romano and George L. Engel, "Delirium: I. Electroencephalographic Data," *Archives of Neurologic Psychiatry* 51 (April 1944): 356–77; Eliere J. Tolan and Fredrich A. Lingel, "'Model Psychosis' Produced by Inhalation of Gasoline Fumes," *American Journal of Psychiatry* 120 (February 1964): 757–61; Eugen Bleuler, *Dementia Praecox, or*

*the Group of Schizphrenias*, trans. J. Zinkin (New York: International Universities Press, 1950); George L. Engel and John Romano, "Delirium: A Syndrome of Cerebral Insufficiency," *Journal of Chronic Disorders* 9 (March 1959): 260–77; Glaser, "Inhalation Psychosis," 315–16.

35. Glaser, "Inhalation Psychosis," 317. See also Ackerly and Gibson, "Lighter Fluid 'Sniffing'"; Clinger and Johnson, "Gasoline Vapors."

36. Glaser, "Inhalation Psychosis," 318.

37. Ibid., 317–18. Michael J. Pescor, "A Statistical Analysis of the Clinical Records of Hospitalized Drug Addicts," Supplement No. 143, *Public Health Reports* (Washington, DC: USGPO, 1943); James V. Lowry, "Hospital Treatment of the Narcotic Addict," *Federal Probation* 20 (December 1956): 42–51; Edwards, "Case Report of Gasoline Sniffing"; J. M. Schneck, "Chloroform Habituation: With a Case Report of Its Occurrence in a Case of Schizophrenia," *Bulletin of the Menninger Clinic* 9 (January 1945): 12–17; W. B. Grant, "Inhalation of Gasoline Fumes by a Child," *Psychiatric Quarterly* 36 (July 1962): 555–57.

38. Glaser, "Inhalation Psychosis," 318–19.

39. Sylvan Bartlett and Fernando Tapia, "Glue and Gasoline 'Sniffing': The Addiction of Youth," *Missouri Medicine* 63 (April 1966): 271.

40. Ibid.

41. Ibid., 270–72.

## Notes to Chapter 5

1. Sokol, "Among Juveniles," 19.

2. William D. Hartley, "Hazards at Home," *Wall Street Journal*, August 20, 1963, 1; Sokol, "Among Juveniles," 18–20. The original provision set a minor's age at twenty-one, while the final law reduced the age to eighteen. At the same time that California began debating a glue-sniffing law in 1963, the Honolulu City Council made sniffing glue illegal, and this was followed soon by Hawaii itself.

3. Fort, *Pleasure Seekers*, 73–75. For more on Kesey and Cal-Berkeley, see Kenneth Keniston, *Young Radicals: Notes on Committed Youth* (New York: Harcourt Brace, 1968); Theodore Roszak, ed., *The Dissenting Academy* (New York: Pantheon, 1968); Tom Wolfe, *The Electric Kool-Aid Acid Test* (1968; New York: Picador, 2008).

4. Sokol, "Among Juveniles," 20.

5. Ibid., 19–20; Verhulst and Crotty, "Glue-Sniffing, II," 3–4; "Stole Glue to Sniff It, Police Told," *Denver Post*, May 18, 1966, Denver Public Library, TN#94111; "Three Men Sentenced for Glue Sniffing," *Denver Post*, June 10, 1966, Denver Public Library, TN#95000.

6. Sokol, "Among Juveniles," 19.

7. Verhulst and Crotty, "Glue-Sniffing, II," 4.

8. Ibid., 4–6; Sokol, "Among Juveniles," 19, 20; Corliss, "Review of the Evidence," 443–44. Leland Corliss reported a case involving a thirteen-year-old leader of a glue-sniffing burglary ring. Among their heists was a group of charm bracelets, one of which included the Ten Commandments, "including, of course, 'Thou Shalt Not Steal.'"

9. "Crime and the Citizen," *Christian Science Monitor*, March 23, 1964, 14; Emilie Tavel, "Gaming, Drugs Interlace Harlem," *Christian Science Monitor*, March 6, 1964, 1, 13.

10. Corliss, "Review of the Evidence," 445.

11. Philip E. Norton, "Perilous Teen-Age Fad, Glue-Sniffing, Alarms Hobby Kit Industry," *Wall Street Journal*, December 7, 1962, 1.
12. Ibid.
13. Ibid.; Hartley, "Hazards at Home," *Wall Street Journal*, August 20, 1963, 1.
14. Norton, "Perilous Teen-Age Fad, Glue-Sniffing, Alarms Hobby Kit Industry," *Wall Street Journal*, December 7, 1962, 1.
15. Martin Arnold, "Murphy Discounts Fears of Violence Here This Summer," *New York Times*, June 9, 1964, 1.
16. "Boy, Thirteen, Sniffing Glue, Falls into Gowanus and Drowns," *New York Times*, August 26, 1964, 16; "Curb on Glue Sales Urged," *New York Times*, August 30, 1964, 94; "Glue-Sniffing Action Asked," *New York Times*, December 25, 1964, 10.
17. "Prisoners Get 'High' on Aromatic Spice," *New York Times*, October 24, 1964, 13; "Girls Riot on Sniffing Polish," *New York Times*, November 14, 1964, 17; "Sniffing of Fluid Is Fatal to Boy; Four Others Made Ill," *New York Times*, June 19, 1965, 25; "Youth Dies after Sniffing Cleaning Fluid at Party," *New York Times*, January 9, 1965, 52.
18. Paul Gardner, "WCBS Will Show Marijuana Party," *New York Times*, February 26, 1965, 59; "Narcotics Agency Formed in Nassau to Fight Addiction," *New York Times*, February 28, 1965, 46.
19. George Lardner Jr., "Glue-Sniffing Fad Fades: So Does Enforcement of Ban," *Washington Post*, March 3, 1965, A18.
20. *New York v. Anonymous* (1965), 260 NYS 2d 860; "New York v. Anonymous (1965)," *New York Supplement*, 2nd Ser., vol. 260 (St. Paul, MN: West Publishing, 1965), 860–62.
21. *New York v. Anonymous* (1965), 260 NYS 2d 860.
22. Cornelius J. Dwyer, "Letter to the Editor: Legal Drinking Age," *New York Times*, March 8, 1965, 28; Morris Kaplan, "City Is a Center in Poison Control," *New York Times*, March 7, 1965, 76.
23. "Glue-Sniffing Study Funded," *Denver Post*, October 30, 1966, 2.
24. "More Money Given Glue-Sniffer Project," *Washington Post*, October 28, 1966, A4; "Glue-Sniffing Study Funded," *Denver Post*, October 30, 1966, 2.
25. In many senses, moralists had "declared war" on sniffing model airplane glue. That kind of melodramatic rhetoric has been shown to oversimplify and distort many social problems where such language has been used and ultimately stunted the growth of legitimate solutions to those problems. See Joel Best, *Random Violence: How We Talk about New Crimes and New Victims* (Berkeley: University of California Press, 1999), in particular Best's chapter on the consequences of "declaring war on social problems," 142–61. Bernard Schissel made a similar argument about stigmatizing Canadian street youth in the 1980s and 1990s in Bernard Schissel, *Blaming Children: Youth Crime, Model Panics, and the Politics of Hate* (Halifax, Nova Scotia: Fernwood Publishing, 1997).
26. David T. Courtwright, *Violent Land: Single Men and Social Disorder from the Frontier to the Inner City* (Cambridge, MA: Harvard University Press, 1996), 228–35.
27. "Police Mobile Unit Opening City Drive on Narcotics Peril," *New York Times*, July 15, 1965, 35.
28. Ibid., 35; "Display Ad 91," *Washington Post*, November 14, 1965, E7.
29. "Jersey Town Irate as Chief of Police Draws Suspension," *New York Times*, June 26, 1965, 31.

30. Martin Tolchin, "Narcotics Traffic Up in Connecticut," *New York Times*, November 23, 1965, 47, 49.

31. Ibid., 49; "The Nation: Trouble in Darien," *New York Times*, November 28, 1965, E2; "Nine Teen-Agers Held," *New York Times*, December 5, 1965, 57.

32. Francis X. Clines, "Narcotics Case Stirs Northport," *New York Times*, December 6, 1965, 45.

33. Francis X. Clines, "L.I. Youths Seized in Narcotics Raid," *New York Times*, December 5, 1965, 57; Clines, "Narcotics Case Stirs Northport," 45.

34. Fort, *Pleasure Seekers*, 97–148.

35. "Two Youths Sterilized by Glue-Sniffing," *Washington Post*, February 6, 1966, A35; Fred M. Hechinger, "Education: Drugs—Threat on Campus," *New York Times*, April 10, 1966, 151.

36. John Hersey, *Too Far to Walk* (New York: Alfred A. Knopf, 1966); Eliot Fremont-Smith, "Some Faust," *New York Times*, February 28, 1966, 25.

37. Fort, *Pleasure Seekers*, 29; Helen Dewar, "Legislators Asked for New Help for Young Offenders," *Washington Post*, December 11, 1965, C40; Thomas A. Johnson, "Harlem Pressing Its Plea for Help," *New York Times*, April 19, 1966, 43.

38. McCandlish Phillips, "Pastor Uses Café as a Lure to Help Teen-Agers," *New York Times*, June 13, 1966, 41.

39. Brill quoted in Evert Clark, "New Aid Foreseen for Drug Addicts," *New York Times*, March 29, 1966, 19. Although Brill's insight was functionally correct, he would remain throughout the decade the kind of alarmist that would keep such communication going (see chapter 9).

40. Patience M. Daltry, "Teen-Age Morals, Parental Alarm," *Christian Science Monitor*, May 27, 1966, 6.

41. Ibid. For books of this kind appearing in 1965 and 1966, see Grace Nies Fletcher, *What's Right with Our Young People* (New York: William Morrow, 1966); Dorothy Gordon, *Who Has the Answer?* (New York: E. P. Dutton, 1965); Thelma C. Purtell, *Tonight Is Too Late* (New York: P. S. Eriksson, 1965); James A. Pike, *Teen-Agers and Sex* (Englewood Cliffs, NJ: Prentice Hall, 1965).

42. Johnson, "Harlem Pressing Its Plea for Help," *New York Times*, April 19, 1966, 44; "More Money Given Glue-Sniffer Project," *Washington Post*, October 28, 1966, A4; Jesse W. Lewis Jr. "Marijuana Use Probed at Western," *Washington Post*, May 7, 1966, A1; "Glue-Sniffing Study Funded," *Denver Post*, October 30, 1966, 2.

43. On January 12, 1967, the first episode of *Dragnet* featured detectives attempting to police an LSD user. Their efforts were to no avail until the government finally regulated the substance. For more on the episode, see "The LSD Story," *Dragnet 1967*, January 12, 1967, Internet Movie Database, http://www.imdb.com/title/tt0565680/. For more on the government's relationship with psychotropic drugs, see Martin A. Lee and Bruce Shlain, *Acid Dreams: The CIA, LSD, and the Sixties Rebellion* (New York: Grove Press, 1986).

44. "Public Law 89–756, Nov. 3, 1966," *United States Statutes at Large, 1966*, vol. 80, pt. 1 (Washington, DC: USGPO, 1967), 1303–5.

## Notes to Chapter 6

1. Verhulst and Crotty, "Glue-Sniffing, II," 2.
2. Corliss, "Review of the Evidence," 443.

3. Ted Rubin, ed., *Presentations of a Conference on Inhalation of Glue Fumes and Other Substance Abuse Practices among Adolescents* (Denver: Denver Juvenile Court, 1967), ii, 1-4.

4. Caroline Jean Acker, *Creating the American Junkie: Addiction Research in the Classic Era of Narcotic Control* (Baltimore: Johns Hopkins University Press, 2002), 2-11, 125-211; Caroline Jean Acker, "Portrait of an Addicted Family: Dynamics of Opiate Addiction in the Early Twentieth Century," in *Altering American Consciousness: The History of Alcohol and Drug Use in the United States, 1800-2000*, ed. Sarah W. Tracy and Caroline Jean Acker (Amherst: University of Massachusetts Press, 2004), 165-81. See also David T. Courtwright, *Dark Paradise: A History of Opiate Addiction in America* (1982; Cambridge, MA: Harvard University Press, 2001), 161-85; Robert Greenfield, *Timothy Leary: A Biography* (New York: Houghton Mifflin, 2006), 201-326; Nancy D. Campbell, *Discovering Addiction: The Science and Politics of Substance Abuse Research* (Ann Arbor: University of Michigan Press, 2007).

5. See H. Ted Rubin Papers, WH2106, Western History Collection, Denver Public Library, Denver, Colorado.

6. University of California Academic Senate, *University of California: In Memoriam, 1969* (Berkeley: University of California, 1969), 92-94.

7. Joseph D. Lohman, "Youth: New Problems and New Directions," in Rubin, *Presentations of a Conference*, 19.

8. Ibid., 22.

9. Ibid., 24.

10. Ibid., 24-25.

11. Ibid., 26.

12. Ibid., 26-27.

13. Ibid., 31.

14. Ibid., 32-33.

15. Ibid., 34-35.

16. William A. Meloff, "Deviant Attitudes and Behavior of Glue-Sniffers in Comparison with Similar and Different Class Peer Groups," in Rubin, *Presentations of a Conference*, 74.

17. Ibid., 75-77.

18. Ibid., 79.

19. Ibid., 77-81.

20. Robert C. Hanson, "Explaining Glue Sniffing and Related Juvenile Delinquency," in Rubin, *Presentations of a Conference*, 84.

21. Hanson, "Explaining Glue Sniffing," 82-87.

22. Ibid., 88.

23. Ibid., 89.

24. Ibid., 87-94.

25. Alan K. Done, "Presentation before Conference on Substance Abuse Practices among Adolescents," in Rubin, *Presentations of a Conference*, 44-45.

26. Ibid., 45-47.

27. Ibid., 47.

28. John William Rawlin, "Identification of the Problem of Substance Abusing Adolescents: The Problem of Amphetamine Abuse," in Rubin, *Presentations of a Conference*, 123.

29. Ibid., 119-21.

30. Victor Gioscia, "Glue Sniffing: Exploratory Hypotheses on the Psychosocial Dynamics of Respiratory Intrajection," in Rubin, *Presentations of a Conference*, 60.

31. Ibid., 63.

216   NOTES TO CHAPTER 7

32. Ibid., 64.
33. Ibid., 66.
34. Ibid., 63–66.
35. Richard Davis, "Report on the Problem of Glue Sniffing in Children and the Work of the New York City Police Department and Its Youth Investigation Bureau in Combating This Problem," in Rubin, *Presentations of a Conference*, 105.
36. Ibid., 104–6.
37. Ibid., 109.
38. Ibid., 109.
39. Dale F. Ely, "Substance Abuse—A Community Problem: The School," in Rubin, *Presentations of a Conference*, 124–25; Milton Luger, "Substance Abuse—A Community Problem," in Rubin, *Presentations of a Conference*, 126–28.
40. Richard Brotman, "Adolescent Substance Use: A Growing Form of Dissent," in Rubin, *Presentations of a Conference*, 37.
41. Ibid., 37–38.
42. Ibid., 42.
43. Ibid., 43.
44. Ibid., 39–43.
45. Ted Alex, "Denver Juvenile Court Glue-Sniffing Project," in Rubin, *Presentations of a Conference*, 96–103.
46. Robert H. Sobolevitch, "Juvenile Substance Abuse in Philadelphia," in Rubin, *Presentations of a Conference*, 112.
47. Ibid., 111–12.
48. Ibid., 116.
49. Harry Silverstein, "Summary of Discussion from Workshop 'A,'" in Rubin, *Presentations of a Conference*, 129–31; Lenore Kupperstein, "Summary of Discussion from Workshop 'B,'" in Rubin, *Presentations of a Conference*, 132–33; Ralph Tefferteller, "Summary of Discussion from Workshop 'C,'" in Rubin, *Presentations of a Conference*, 134–35; Roberta Wilson, "Summary of Discussion from Workshop 'D,'" in Rubin, *Presentations of a Conference*, 136–38.
50. Ted Rubin, "An Appeal for Early and Intensive Intervention," in Rubin, *Presentations of a Conference*, 139.
51. Ibid., 139–40.
52. Ibid.

## Notes to Chapter 7

1. "Whiff of Innocence," *Time*, August 25, 1967, 51; also "Glue-Sniffer, Fourteen, Charged in Killing of Two Girls," *Washington Post*, May 1, 1967, A3; "Glue Sniffer Freed in Sex Slaying of Two," *Washington Post*, August 17, 1967, A2.
2. Charles L. Winek, Wellon D. Collom, and Cyril H. Wecht, "Fatal Benzene Exposure by Glue-Sniffing," *The Lancet*, March 25, 1967, 683.
3. Julius Merry, "Glue Sniffing and Heroin Abuse," *British Medical Journal* 2 (May 6, 1967): 360.
4. Merry, "Glue Sniffing and Heroin Abuse," 360. Merry was writing in Britain, but he was talking about the United States. Although his exemplary case was an Englishman,

Merry and others admitted that glue sniffing had yet to become a real problem in Britain, but its explosion in the United States required British doctors and law enforcement officers to be on guard. See also M. M. Glatt, "Drug Treatment Centres," *British Medical Journal* 3 (July 22, 1967): 242.

5. Lester G. Thomas, "The Denver Glue Sniffing Project," *Juvenile Court Judges Journal* 18 (July 1967): 46.

6. Acker, *Creating the American Junkie*, 125-211. Thomas, "Denver Glue Sniffing Project," 46-47. It was yet another example of the problem described by Caroline Jean Acker (see chapter 6).

7. Thomas, "Denver Glue Sniffing Project," 47-48. Thomas also outlined the prevention measures presented at the glue summit and closed the article with treatment ideas that had already appeared elsewhere.

8. Courtwright, *Dark Paradise*, 161-85; Acker, *Creating the American Junkie*, 125-211.

9. Edward Press and Alan K. Done, "Solvent Sniffing: Physiologic Effects and Community Control Measures for Intoxication from the Intentional Inhalation of Organic Solvents, I," *Pediatrics* 39 (March 1967): 451.

10. Ibid., 451, 452.

11. Ibid., 452-54.

12. Ibid., 458.

13. Ibid., 454-59.

14. Edward Press and Alan K. Done, "Solvent Sniffing: Physiologic Effects and Community Control Measures for Intoxication from the Intentional Inhalation of Organic Solvents, II," *Pediatrics* 39 (April 1967): 611.

15. Ibid., 611-17; G. Christiansson and B. Karlsson, "Sniffing, Method of Intoxication among Children," *Svensk Lakartidn* 54 (January 1957): 33.

16. Press and Done, "Solvent Sniffing: II," 617.

17. Ibid.

18. Ibid., 618.

19. Ibid., 613-17, 618.

20. Ibid., 618-19.

21. Ibid., 619-20.

22. Martin H. Keeler and Clifford B. Reifler, "The Occurrence of Glue Sniffing on a University Campus," *Journal of the American College Health Association* 16 (October 1967): 69.

23. Ibid.

24. Ibid., 70; Martin H. Keeler, "The Use of Hyoscyamine as a Hallucinogen and Intoxicant," *American Journal of Psychiatry* 124 (December 6, 1967): 852-54. These weren't the only problems. Another eighteen-year-old, this one a high school dropout who had experimented with model airplane glue in the past, developed a dependence on Asthmador, a mixture of belladonna and stramonium, the principal alkaloid of both being hyoscyamine. It produced hallucinogenic effects comparable to peyote.

25. J. Robertson Unwin, "Illicit Drug Use among Canadian Youth: Part I," *Canadian Medical Association Journal* 98 (February 24, 1968): 403.

26. Ibid.

27. Ibid., 402-3; Vera Gellman, "Glue-Sniffing among Winnipeg School Children," *Canadian Medical Association Journal* 98 (February 24, 1968): 411.

28. Gellman, "Winnipeg School Children," 411.

29. Ibid., 411-12.

30. States passing legislation regulating model airplane glue through 1967 were California, Connecticut, Florida, Illinois, Louisiana, Maine, Maryland, Massachusetts, Michigan, New Jersey, New York, and Rhode Island. For a more detailed discussion of state laws and the processes by which states passed them, see chapter 8.

31. Lenore R. Kupperstein and Ralph M. Susman, "A Bibliography on the Inhalation of Glue Fumes and Other Toxic Vapors—A Substance Abuse Practice among Adolescents," *International Journal of the Addictions* 3 (Spring 1968): 177–78. See also Louis J. Ravin and Peter D. Bernardo, "Pharmaceutical Sciences—1967: A Literature Review of Pharmaceutics," *Journal of Pharmaceutical Sciences* 57 (July 1968): 1075–97; Gordon E. Barnes and Brent A. Vulcano, "Bibliography of the Solvent Abuse Literature," *International Journal of the Addictions* 14, no. 3 (1979): 401–21.

32. Kupperstein and Susman, "A Bibliography," 177–97. The bibliography was fraught with errors and sometimes mysterious seemingly nonexistent citations, but it was a start, if nothing else.

33. James L. Chapel and Daniel W. Taylor, "Glue Sniffing," *Missouri Medicine* 65 (April 1965): 288–89.

34. Ibid., 291.

35. Ibid., 289–92.

36. Charles L. Winek, Cyril H. Wecht, and Wellon D. Collom, "Toluene Fatality from Glue Sniffing," *Pennsylvania Medicine* 71 (April 1968): 81.

37. Mack I. Shanholtz, "Glue Sniffing: A New Symptom of an Old Disease," *Virginia Medical Monthly* 95 (May 1968): 304–5.

38. Pamela Hinton and Bernd Koch, "Hemolytic Uremic Syndrome: Report of Two Cases," *Canadian Medical Association Journal* 98 (April 27, 1968): 819–23.

39. John Todd, "'Sniffing' and Addiction," *British Medical Journal* 4 (October 26, 1968): 255–56.

40. Robert W. Deisher, Albert J. Schroeder, V. Robert Allen, Harry Bakwin, Victor Eisner, Dale C. Garell, S. L. Hammar, Sprague W. Hazard, Thomas E. Shaffer, John Allen Welty, Charles Louis Wood, Charles Keck, and Graham Blaine, "Drug Abuse in Adolescence: The Use of Harmful Drugs, a Pediatric Concern," *Pediatrics* 44 (July 1969): 133–34.

41. Henry L. Verhulst and John J. Crotty, "Glue-Sniffing Deterrent," *Bulletin: National Clearinghouse for Poison Control Centers* (November–December 1969): 5.

42. Ibid., 4.

## Notes to Chapter 8

1. "Glue Sniffing Could Become More Serious," *Baton Rouge Advocate*, April 21, 1966, 4C.

2. Kupperstein and Susman, "A Bibliography," 183. The states that had passed such laws were California, Connecticut, Illinois, Maine, Maryland, Massachusetts, Michigan, New Jersey, New York, and Rhode Island. See also Fort, *Pleasure Seekers*, 73–78.

3. "Glue Sniffing Could Become More Serious," *Baton Rouge Advocate*, April 21, 1966, 4C.

4. "House Bill No. 752," *Legislative Calendar of the State of Louisiana*, Twenty-Ninth Regular Session of the Legislature, 1966, 193; "House Bill No. 752," *Official Journal of the*

*Proceedings of the House of Representatives of the State of Louisiana*, Twenty-Ninth Regular Session of the Legislature, 1966, 288, 330, 643, 685, 907-8.

5. "Model Glue Bill Sent to Floor by House Unit," *Baton Rouge Advocate*, June 3, 1966, 10C.

6. "Monroe Student Found Dead in Abandoned Car," *Baton Rouge Advocate*, June 11, 1966, 10A; "Gas Sniffing Blamed Here as Boy Dies," *Monroe Morning World*, June 11, 1966, 1.

7. "Monroe Student Found Dead in Abandoned Car," *Baton Rouge Advocate*, June 11, 1966, 10A.

8. "Glue Sniffing Is Outlawed by N.O. Council," *Baton Rouge Advocate*, June 24, 1966, 15A; "City Council Passes Law to Curb 'Glue Sniffing,'" *Louisiana Weekly*, July 2, 1966, 2, 8.

9. "City Council Passes Law to Curb 'Glue Sniffing,'" *Louisiana Weekly*, July 2, 1966, 8.

10. David R. Poynter, *Membership in the Louisiana House of Representatives, 1812-2012* (Baton Rouge: Louisiana House of Representatives, 2010), 10, 12, 22, 33, 44, 54, 81, 93, 110, 116, 126, 135, 196, 225, 232, 235, 276, 286, 290, 293. The bill's sponsors (with their parishes) were Eddie L. Sapir, William A. Gill Jr., Donald L. Fortier, Edward L. Boesch, Charles Smither, Ernest J. Hessler Jr., Charles Bordes III, Harry J. Hillensbeck, Joseph S. Casey, Clyde F. Bell Jr., Edward F. LeBreton Jr., John P. Sullivan, Eugene G. O'Brien, Arthur A. Crais, Thomas A. Early Jr., Stephen K. Daley, Vernon J. Gregson, Salvador Anzelmo, and Anthony J. Vesich Jr. (Orleans Parish); James E. Beeson, William J. Dwyer, and Francis E. "Hank" Lauricella (Jefferson Parish); Luther F. Cole, Joe Keogh, Lillian W. Walker, and William F. "Bill" Bernhard Jr. (East Baton Rouge Parish); J. Bennett Johnston Jr., and Taylor W. O'Hearn (R) (Caddo Parish); Roderick L. "Rod" Miller (R) (Lafayette Parish); Richard E. Talbot and Cleveland J. Marcel Sr. (Terrebonne Parish); Joel T. Chaisson (St. Charles Parish); Risley C. "Pappy" Triche (Assumption Parish); Allen C. Gremillion (Acadia Parish); S. M. Morgan Jr. (Red River Parish); Gordon E. "Buddy" Causey (Tangipahoa Parish); W. J. "Edge" Richardson (Caldwell Parish); T. J. Strother (Allen Parish); Richard P. "Dick" Guidry (Lafourche Parish); Joseph Emile Coreil (Evangeline Parish); Thomas Marx Hoffman (Iberville Parish); Samuel B. Nunez Jr. (St. Bernard Parish); J. L. Lacy (Bienville Parish); James P. Smith (Union Parish); Warren J. Simon (Vermilion Parish).

11. "House Bill No. 752," *Official Journal of the Proceedings of the House of Representatives of the State of Louisiana*, 907-8; Poynter, *Membership*, 49, 55, 73, 222. The four representatives voting against the original bill were Joe Henry Cooper from DeSoto Parish, Conway LeBleu from Cameron Parish, Larry Parker from Rapides Parish, and Harry M. Hollins from Calcasieu Parish.

12. "UVL Backs Legislation to Stop 'Glue Sniffing,'" *Louisiana Weekly*, June 11, 1966, 10; Adam Fairclough, *Race and Democracy: The Civil Rights Struggle in Louisiana, 1915-1972* (Athens: University of Georgia Press, 1995), 513.

13. Jules G. Mollere was the state senator from Jefferson Parish, west New Orleans. "House Bill No. 752," *Legislative Calendar of the State of Louisiana*, 193; "House Bill No. 752," *Official Journal of the Proceedings of the Senate of the State of Louisiana*, Twenty-Ninth Regular Session of the Legislature, 1966, 599, 601, 1403-4; "US Judge Orders a New Trial for Head of Louisiana Senate," *New York Times*, October 28, 1982, 28; "Louisiana State Senate President Is Found Guilty," *Spokesman-Review*, February 5, 1983, 36; and Arthur E. McEnany, *Membership in the Louisiana Senate, 1880-2004* (Baton Rouge: Louisiana State Senate, 2002), 60, 86-87. The others were Michael O'Keefe, Olaf Fink, and Jules Mollere.

Michael H. O'Keefe was also from New Orleans. He later became president of the Louisiana Senate, a post he used as a staging ground for mail fraud and obstruction of justice. He was convicted in 1983. Olaf J. Fink was yet another New Orleans politician, serving as senator from Orleans Parish.

14. "Act No. 110 (House Bill No. 752)," State of Louisiana, *Acts of the Legislature*, vol. 1, regular session, 1966, 306–7. Substances specifically banned in the legislation were toluol, hexane, trichloroethylene, acetone, toluene, ethyl acetate, methyl ethyl ketone, trichlorochthane, isopropanol, methyl isobutyl ketone, methyl cellosolve acetate, and cyclohexanone.

15. Ibid.

16. "House Bill No. 752," *Official Journal of the Proceedings of the Senate of the State of Louisiana*, 1558. The number was actually higher than twenty, because senate revisions placed, for example, toluol and toluene in the same category and worded other entries differently. The list in the proposed amendment included acetone, amyl acetate, benzol or benzene, butyl acetate, butyl alcohol, carbon tetrachloride, chloroform, cyclohexanone, ethanol or ethyl alcohol, ethyl acetate, hexane, isopropanol or isopropyl alcohol, isopropyl acetate, methyl "cellosolve" acetate, methyl ethyl ketone, methyl isobutyl ketone, toluol or toluene, trichloroethylene, tricresyl phosphate, and xylol or xylene.

17. Ibid. The twenty-four-hour sales ban did provide an exception for model or hobby kits that included glue in their packaging.

18. Ibid., 1509, 1513–14, 1558–59, 1593–94; "House Bill No. 752," *Legislative Calendar of the State of Louisiana*, 193; "House Bill No. 752," *Official Journal of the Proceedings of the House of Representatives of the State of Louisiana*, 1884, 1886. Cooper, LeBleu, and Hollins all voted for the revised final version of the bill. Parker was absent.

19. "Criminal Code: Part V. Offenses Affecting the Public Morals," *Louisiana Revised Statutes of 1950, Act 2 of the Extraordinary Session of 1950*, vol. 2 (St. Paul, MN: West Publishing, 1950), 369–73; "Offenses Affecting the Health and Morals of Minors," *West's Louisiana Statutes Annotated: Revised Statutes, Sections 14:1 to 14:End*, vol. 9 (St. Paul, MN: West Publishing, 1951), 531–38.

20. "Criminal Code: Part V. Offenses Affecting the Public Morals," *Louisiana Revised Statutes 1972*, Pocket Parts, vol. 2 (Baton Rouge: Louisiana State Law Institute, 1973), 146.

21. Ibid., 145–51.

22. *Louisiana v. Dimopoullas* (1972), 260 La. 874, 257 So. 2d 644; "Criminal Code: § 93.1. Model glue; use of; unlawful sales to minors; penalties," *West's Louisiana Statutes Annotated: Revised Statutes: Sections 14:74 to 14:End*, vol. 9A (St. Paul, MN: West Publishing, 1986), 120–21.

23. "HB 1378," *Journal of the House of Representatives*, vol. 1, Regular Session [Including Extension] of the Forty-First Legislature [under the Constitution of 1885], April 4, 1967, through July 14, 1967, 406; "SB 893," *Journal of the Senate, State of Florida*, Forty-First Regular Session April 4 through July 14, 1967, 294.

24. "SB 893," *Journal of the Senate, State of Florida*, 294; "Members of the Senate, Regular Session, 1967," *Journal of the Senate, State of Florida*, Forty-First Regular Session April 4 through July 14, 1967, 24. The bill's cosponsors were Democrats John R. Broxon (Gulf Breeze), J. Emory Cross (Gainesville), John E. Mathews Jr. (Jacksonville), Verle A. Pope (St. Augustine), T. Truett Ott (Tampa), Ben Hill Griffin Jr. (Frostproof), Jerry Thomas (Lake Park), Robert L. Shevin (Miami), George L. Hollahan Jr. (Coral Gables), Ralph R. Poston (Miami), Richard B. Stone (Miami), and Republicans Elizabeth Johnson (Cocoa

Beach), Ralph R. Clayton (DeLand), Kenneth Plante (Altamonte Springs), Robert H. Helrod (Windermere), C. W. Bill Young (Pinellas Park), Harold S. Wilson (Bellair), Richard J. Deeb (St. Petersburg), Joseph A. McClain Jr. (Tampa), Warren S. Henderson (Venice), David C. Lane (Ft. Lauderdale), and John W. Bell (Ft. Lauderdale).

25. "HB 1378," *Journal of the House of Representatives*, vol. 1, 406, 667, 785, 1030, 1679; "Members of the House of Representatives, Regular Session 1967," *Journal of the House of Representatives*, vol. 1, Regular Session [Including Extension] of the Forty-First Legislature [under the Constitution of 1885], April 4, 1967, through July 14, 1967, 9. Joining Firestone was Harold G. Featherstone, a Democrat from nearby Hialeah, and Guy W. Spicola, a Democrat from Tampa, as well as Robert W. Rust, a Republican from Palm Beach, and Thomas M. Gallen, a Democrat from Bradenton.

26. "SB 893," *Journal of the Senate, State of Florida*, 747, 1111, 1133, 1161, 1206.

27. Ibid., 1257, 1292.

28. "HB 1378," *Journal of the House of Representatives*, vol. 1, 406, 941; "HB 1378," *Journal of the House of Representatives*, vol. 2, Regular Session [Including Extension] of the Forty-First Legislature [under the Constitution of 1885], April 4, 1967, through July 14, 1967, 1313, 1340, 1369, 1454.

29. *History of Legislation, 1967 Regular Session, Florida Legislature*, Legislative Information Division, Joint Legislative Management Committee, Photocopy Reduction (Jacksonville: Fla-Ga Law Publishing, 1988), 173, 255; "HB 1378," *Journal of the House of Representatives*, vol. 2, 1682, 1714–15; "SB 893," *Journal of the Senate, State of Florida*, 1257, 1292, 1676.

30. "Chapter 67-416; Senate Bill No. 893," *Regular Session 1967, General Acts and Resolutions Adopted by the Legislature of Florida*, vol. 1, pt. 1, 1967, 1286–88. Florida's list of banned substances included acetone, amyl acetate, benzol or benzene, butyl acetate, butyl alcohol, carbon tetrachloride, chloroform, cyclohexanone, ethanol or ethyl alcohol, ethyl acetate, hexane, isopropanol or isopropyl alcohol, isopropyl acetate, methyl "cellosolve" acetate, methyl ethyl ketone, methyl isobutyl ketone, toluol or toluene, trichloroethylene, tricresyl phosphate, and xylol or sylene.

31. "Chapter 877: Miscellaneous Crimes," *Official Florida Statutes 1967*, State of Florida, 1967, 3843–46; "877.11, Inhalation, ingestion, possession, sale, purchase, or transfer of harmful chemical substances; penalties," *West's Florida Statutes Annotated*, vol. 22D, Title XLVI, Crimes (Egan, MN: West Group, 2000), 478–80.

32. *Florida Juvenile Court Statistics, 1965, 1966, 1967* (Tallahassee: State Department of Public Welfare, 1968), Florida State Library, iii, v, xv, xvii, xxvii, xxix.

33. "Drugs and Substance Abuse," *Florida Health Notes* 61 (October 1969): 272. See also Department of Education, *Status Report: Drug Education, 1971–72* (Tallahassee: State of Florida, January 1972), 1; *Drug Abuse in Florida* (Tallahassee: Governor's Task Force on Narcotics, May 1970), 1–10; *1974 Summary of Florida State Plan for Drug Abuse Prevention* (Tallahassee: Florida Department of Health and Rehabilitative Services, 1974), Florida State Library, 4–19.

34. *In the Interest of P.G. and G.G., Minors*, No. 73–47, District Court of Appeal of Florida, Third District, July 3, 1973; *H.R.H., a juvenile, Appellant, v. The State of Florida, Appellee*, No. 74–420, District Court of Appeal of Florida, Third District, October 29, 1974.

35. Earl Faircloth, "Crimes: Inhalation or Possession of Model Glue," *Biennial Report of the Attorney General, State of Florida* (Tallahassee, 1969), 242; *Linville v. Florida*, No. 51583, Supreme Court of Florida, May 18, 1978.

36. State Department of Public Welfare, *Mississippi Youth Court Statistics* (Jackson: Mississippi Department of Public Welfare, 1961–1966); ibid., (Jackson: Mississippi Department of Public Welfare, 1967), 16, 17.

37. Ibid., 4, 16, 17; ibid., (Jackson: Mississippi Department of Public Welfare, 1968), 17.

38. "HB No. 281," *Journal of the House of Representatives of the State of Mississippi* (Jackson: Hederman Brothers, 1968), 126, 328–29, 345, 352–53, 364, 1056, 1068–69, 1080; "SB No. 1702," *Journal of the House of Representatives of the State of Mississippi* (Jackson: Hederman Brothers, 1968), 345; "HB No. 281," *Journal of the Senate of the State of Mississippi* (Jackson: Hederman Brothers, 1968), 130–31, 267, 273–74, 289, 334, 938–39, 1092–93.

39. "Part XIII: Biographical Data of Senators, 1968–1972," *Journal of the Senate of the State of Mississippi* (Jackson: Hederman Brothers, 1968), 1777; "SB No. 1702," *Journal of the Senate of the State of Mississippi* (Jackson: Hederman Brothers, 1968), 131. William E. McKinley and Jean D. Muirhead (listed as Mrs. Marvin L. Muirhead) both represented Mississippi's twenty-seventh senatorial district, Hinds County.

40. "Part XIII: Biographical Data of Senators, 1968–1972," *Journal of the Senate of the State of Mississippi* (Jackson: Hederman Brothers, 1968), 1774–86; "SB No. 1702," *Journal of the Senate of the State of Mississippi*, 273. William Valentine Jones, from Waynesboro, was both an attorney and a cattle farmer. Corbet Lee Patridge was from Schalter, in Leflore County, a cotton farmer who had served as public relations representative for the National Cotton Council. Theodore Smith, from Corinth, was another farmer who represented both Alcorn and Tippah Counties. Thomas Arnie Watson, from North Carrollton represented Attala, Carroll, and Montgomery Counties. He was a cattle and tree farmer. The fifth no vote was Biloxi's John Thomas Munro, and Biloxi, though rural by some standards, was the seat of Harrison County, the state's second most populous county. Louisiana had two Republicans cosponsoring legislation, but that was largely a creature of large urban areas carrying the capacity for a handful of Republican legislators. Mississippi had no such Republicans. Its entire roster of legislators were Democrats.

41. "House of Representatives by Districts and Counties," *Journal of the House of Representatives of the State of Mississippi* (Jackson: Hederman Brothers, 1968), 1525–43; "HB No. 281," *Journal of the House of Representatives of the State of Mississippi*, 126. Russell Davis, Sutton Marks, Ralph Sowell, Joseph Moss, Charles Mitchell, and Emmett Owens all represented Jackson's Hinds County. James B. True, Tommy Gollott, Clyde Woodfield, and James C. Simpson were from Harrison County, home to Biloxi and Gulfport in the far southern portion of the state. Ralph Herrin was from Collins, Frank Carlton from Greenville, Marvin B. Henley from Philadelphia, Hubert Noel Finnie from Courtland, and Helen McDade from DeKalb.

42. "House of Representatives by Districts and Counties," *Journal of the House of Representatives of the State of Mississippi*, 1525–43; "HB No. 281," ibid., 353.

43. "Part XIII: Biographical Data of Senators, 1968–1972," *Journal of the Senate of the State of Mississippi*, 1774–86; "HB No. 281," *Journal of the Senate of the State of Mississippi*, 1093. The senators voting no on the final version of Mississippi's bill were Munroe; William B. Alexander, a lawyer from Cleveland; William Ervin Corr, a lawyer from Sardis; Joseph McRae Mosby, a farmer and meat packer from urban Meridian; John William Powell, a cattle and timber farmer from Liberty; R. B. "Breezy" Reeves, an attorney from McComb; Kenneth Barkley Robertson, an attorney from Pascagoula; Ben Harry Stone, an attorney from Gulfport in populous Harrison County; and H. C. Strider, a farmer from Charleston.

NOTES TO CHAPTER 8    223

44. "Chapter 347: House Bill No. 281," *Laws of the State of Mississippi*, 1968, 483; "Act No. 110 (House Bill No. 752)," State of Louisiana, *Acts of the Legislature*, 306–7.

45. "Crimes and Misdemeanors, Chapter 1, Title 11," *Mississippi Code 1942 Annotated, Recompiled*, vol. 2A—*1956, Crimes; Special Actions* (Atlanta: Harrison Company, 1957), 216–19; "Crimes Affecting Public Health: Poisons; Inhalation of Toxic Vapors from Model Glue; Unlawful Glue Sales to Minors," *Miss. Code Ann.* § 97-27-33 (1942); "Crimes and Misdemeanors, Chapter 1, Title 11," *1972 Cumulative Supplement to Mississippi Code 1942 Annotated, Recompiled*, vol. 2A (Atlanta: Harrison Company, 1972), 120. For more on governmental regulations as they relate to poisons, see Marian Moser Jones, "Poison Politics: A Contentious History of Consumer Protection against Dangerous Household Chemicals in the United States," *American Journal of Public Health* 103 (May 2013): 801–12.

46. State Department of Public Welfare, *Mississippi Youth Court Statistics* (Jackson: Mississippi Department of Public Welfare, 1961), 17. Of the fifteen delinquent teenagers, the race numbers proved more balanced, and the gender numbers seemed to be moving in that direction. Eight of the defendants were black, seven white, twelve were male, and three were female.

47. Robert Coram, "Glue Sniffers Tempt Injury, Death," *Atlanta Journal and Constitution*, June 12, 1966, 51.

48. Ibid.

49. Ibid.

50. Ibid.

51. Ibid.

52. Ibid.; "A Question Arises," *Atlanta Journal*, July 22, 1968, 18.

53. "'Glue Sniff' Bill Passed in DeKalb," *Atlanta Journal*, August 24, 1966, 82.

54. Junie Hamilton, "City Limits Sale of Glue for Models," *Atlanta Journal*, February 7, 1967, 5.

55. "Glue Sniffing . . . The Quest for Ecstacy," *Georgia's Health* 47 (February 1967): 1. Later that year, the same newsletter reported on the related consequences of sniffing the contents of aerosol cans. See "'Sniffing' Contents of Aerosol Cans Leads to Death," *Georgia's Health* 47 (October 1967): 3.

56. "Quest for Ecstacy," 2; John Askins, "Glue Sniffer Learns Lesson," *Atlanta Journal*, September 21, 1967, 22A.

57. "SB 205," *Journal of the Senate of the State of Georgia at the Regular Session* (Monday, January 8, 1968—Friday, March 8, 1968), 15–16, 36; Orville Gaines, "Boy in Gun Battle Cites Glue Sniffing," *Atlanta Journal*, January 16, 1968, 8B.

58. "SB 205," *Journal of the Senate of the State of Georgia at the Regular Session*, 151, 191–92; "Senate OK's Bill to Outlaw Glue-Sniffing," *Atlanta Journal*, January 19, 1968, 2A; "SB 205," *Journal of the House of Representatives of the State of Georgia at Regular Session* (Monday, January 8, 1968—Friday, March 8, 1968), 2975–76; "Intentional Inhaling of Fumes of Model Glue, etc., No. 1122 (Senate Bill No. 205)," *Acts and Resolutions of the General Assembly of the State of Georgia*, 1968, 1194–96.

59. "SB 205," *Journal of the Senate of the State of Georgia at the Regular Session*, 191–92; "Senate OK's Bill to Outlaw Glue-Sniffing," *Atlanta Journal*, January 19, 1968, 2A.

60. "SB 205," *Journal of the House of Representatives of the State of Georgia at Regular Session*, 499, 511, 603, 1165, 2975–76.

61. "SB 205," *Journal of the House of Representatives of the State of Georgia at Regular Session*, 2975–78, 3404–5; "SB 205," *Journal of the Senate of the State of Georgia at the Regular*

Session, 1800, 1863–64, 2389–92; "Lowery, Starnes Qualify to Run for New Terms," *Rome News-Tribune*, May 24, 1966, 1; "Silent Partners: The Role of the Church in Liberalizing Georgia's Abortion Laws," *Georgia Right to Life*, http://www.grtl.org/history.asp; "Intentional Inhaling of Fumes of Model Glue, etc., No. 1122 (Senate Bill No. 205);" 1194–96; Ben W. Fortson, *Georgia's Official Register, 1967–1968* (Atlanta: Department of Archives and History), 1967, 409–10; *Bond v. Floyd* (1966), 385 US 116. Charles Edward Graves was the other no vote in the Georgia House. He, too, was from Rome's Floyd County and sat next to Starnes in the chamber. The black legislators who approved the measure were there because of the ruling in *Bond v. Floyd* (1966), instigated by civil rights leader Julian Bond, which decided the case on First Amendment protections of free speech.

62. Arthur K. Bolton, "Unofficial Opinion U70-59 (4/7/70)," *Opinions of the Attorney General, 1970* (Atlanta: Harrison Company, 1970), 299. At the time, Bolton cited Article I, Section IV, Paragraph I of the Georgia Constitution: "Laws of a general nature shall have uniform operation throughout this state and no local or special law shall be enacted in any case for which provision has been made by an existing general law." That paragraph has since been moved to Article III, Section VI, Paragraph IV (a), and the legislature has since amended the restriction to include an exception that it may "by general law authorize local governments by local ordinance or resolution to exercise police powers which do not conflict with general laws." It has, in other words, superseded the ruling of both the supreme court's drinking and driving rulings and the attorney general's glue-sniffing opinion to make such allowances acceptable. See also *Giles v. Gibson* (1952), *Reports of Cases Decided in the Supreme Court of the State of Georgia*, April and September Terms 1951, and January Term, 1952, vol. 208 (Atlanta: Harrison Company, 1952), 850–53; *Jenkins, Chief of Police, etc. v. Jones* (1953), *Reports of Cases Decided in the Supreme Court of the State of Georgia*, April and September Terms, 1952, and January and April Terms, 1953, vol. 209 (Atlanta: Harrison Company, 1953), 758–68; *Constitution of the State of Georgia* (Atlanta: Secretary of State, 2009).

63. "Intentional Inhaling of Fumes of Model Glue, etc., No. 1122 (Senate Bill No. 205)," 1194–96; "Controlled Substances: Article 4: Sale, Possession, Transfer, or Inhalation of Model Glue," T16, C13, A4, *Official Code of Georgia Annotated*, vol. 14A, 2007 ed., *Title 16, Crimes and Offenses*, chs. 10–17. Substances banned by the final version of the Georgia act were acetone, amyl chloride (iso- and tertiary), benzene, carbon disulfide, carbon tetrachloride, chloroform, ether, ethyl acetate, ethyl alcohol, ethylene dichloride, isopropyl acetate, isopropyl alcohol, isopropyl ether, methyl acetate, methyl alcohol, propylene dichloride, propylene oxide, trichloroethylene, amyl acetate, amyl alcohol, butyl acetate, butyl alcohol, butyl ether, diethylcarbonate, diethylene oxide (Dioxan), dipropyl ketone, ethyl butyrate, ethylene glycol monoethyl ether (Cellosolve), ethylene glycol monomethyl ether acetate (Methyl Cellosolve Acetate), isobutyl alcohol, methyl amyl acetate, methyl amyl alcohol, methyl isobutyl ketone, and toluene.

64. "Controlled Substances: Article 4: Sale, Possession, Transfer, or Inhalation of Model Glue."

65. "Retiring Chief Scolds Present Generation," *Birmingham News*, May 6, 1969, 14; "HB 200," *Journal of the House of Representatives of the State of Alabama*, Regular Session of 1969, vol. 1, 57.

66. "H 200," *Journal of the House of Representatives of the State of Alabama*, 370, 431–32, 976–77; Dan Dowe, "House Committee Approves School Bus Safety Proposal," *Birmingham News*, June 11, 1969, 12; "Report of Standing Committee, HB 200," Alabama

Secretary of State, Committee Files of the Legislature, Alabama Department of Archives and History, Montgomery, Alabama.

67. "Roster of the House of Representatives of Alabama," *Alabama Laws (and Joint Resolutions) of the Legislature of Alabama, 1969*, vol. 3, 2390-95. W. E. (Bill) Owens Jr., was from Gadsden, Fred Ray Lybrand from Anniston, and Joe C. McCorquodale Jr., from Jackson.

68. Don F. Wasson, "Senate Unit OKs 43 of 87 Bills," *Montgomery Advertiser*, June 11, 1969, 11.

69. "A Bill to Be Entitled an Act," #2-288:5/23/68, Alabama Department of Archives and History, Montgomery, Alabama. The proposed Alabama statute banned acetone, amylacetate, benzol or benzene, butyl acetate, butyl alcohol, carbon tetrachloride, chloroform, cyclohexanone, ethanol or ethyl alcohol, ethyl acetate, hexane, isopropanol or isopropyl alcohol, isopropyl acetate, methyl "Cellosolve" acetate, methyl ethyl ketone, methyl isobutyl ketone, toluol or toluene, trichloroethylene, tricresyl phosphate, xylol or sylene. It also included a ban for "any other solvent, material, substance, chemical or combination thereof, having the property of releasing toxic vapors."

70. Stephen G. Katsinas, "George C. Wallace and the Founding of Alabama's Public Two-Year Colleges," *Journal of Higher Education* 65 (July–August 1994): 451, 454; "Bill Would OK City Building of Park Units," *Birmingham News*, May 7, 1969, 2.

71. Dan Dowe, "Frosh Lawmaker Challenges Fite," *Birmingham News*, June 11, 1969, 59; Dan Dowe, "House Votes Salons in on Paid-Up Insurance," ibid., June 13, 1969, 1, 4; Dan Dowe, "Fite Survives Challenge to Rule in House," ibid., June 13, 1969, 2.

72. Musto, *American Disease*, 244-53; King County Bar Association, "Drugs and the Drug Laws," 27-28.

73. Joseph F. Spillane, "Federal Policy in the Post-Anslinger Era: A Guide to Sources, 1962-2001," in *Federal Drug Control: The Evolution of Policy and Practice*, ed. Jonathon Erlen and Joseph F. Spillane (Binghamton, NY: Haworth Press, 2004), 209-16; Musto, *American Disease*, 251-65; Federal Bureau of Investigation, *Crime in the United States: Uniform Crime Reports, 1991* (Washington, DC: Uniform Crime Reporting Program, 1992), 286-87; US Department of Justice, *Sourcebook of Criminal Statistics, 1990* (Rockville, MD: Justice Statistics Clearinghouse, 1991), 494-95.

74. Barry Glassner, *The Culture of Fear: Why Americans Are Afraid of the Wrong Things* (New York: Basic Books, 1999), xxv; Philip Jenkins, *Moral Panic: Changing Concepts of the Child Molester in Modern America* (New Haven: Yale University Press, 1998), 116-19. In relation to Glassner's comments and the role that legislative decision-making plays in directing moral panics and sapping resources from elsewhere, see also Sheldon Ungar, "Moral Panic versus the Risk Society: The Implications of the Changing Sites of Social Anxiety," *British Journal of Sociology* 52 (June 2001): 271-91. In relation to child sexual abuse as a comparative legislative model for the glue-sniffing epidemic, see also Jeffrey S. Victor, "Moral Panics and the Social Construction of Deviant Behavior: A Theory and Application to the Case of Ritual Child Abuse," *Sociological Perspectives* 41, no. 3 (1998): 541-65.

## Notes to Chapter 9

1. John Carmody, "The District Line," *Washington Post*, June 30, 1967, B8.

2. Mary Kelly, "New York City Schools Face Up to Low Reading Scores," *Christian Science Monitor*, February 1, 1967, 1.

3. Roy R. Silver, "8 Percent of Great Neck High Students Admit They've Used Marijuana," *New York Times*, February 17, 1967, 36; "10 Percent of Pupils in Survey Used Drugs," *Washington Post*, February 17, 1967, A19.

4. Hank Burchard, "Glue-Sniffing Gang Revealed in Fairfax," *Washington Post*, February 21, 1967, C9; Ted L. Rothstein, "Gluesniffing," *Washington Post*, March 24, 1967, A20.

5. Louis Cassels, "Book Gives Facts to Fight Drug Abuse," *Washington Post*, April 19, 1967, A23.

6. Smith, Kline and French Laboratories, *Drug Abuse: Escape to Nowhere* (Washington, DC: National Education Association, 1967). See also Cassels, "Book Gives Facts," A23; Dorothy Rich, "Teach Facts about Drugs," *Washington Post*, May 21, 1967, K14.

7. Rich, "Teach Facts about Drugs," *Washington Post*, May 21, 1967, K14. See also *What We Can Do about Drug Abuse* (New York: Public Affairs Pamphlets, 1967); *Runningawayness* (Washington, DC: USGPO, 1967).

8. Felix Kessler, "Light Up a Banana: Students Bake Peels to Kick Up Their Heels," *Wall Street Journal*, March 30, 1967, 1. See also "Lacquer-Thinner Sniffing Blamed in Five Deaths in Japan," *New York Times*, June 15, 1967, 4; "Two Patrolmen Slay Youth in Brooklyn after a Struggle," *New York Times*, June 25, 1967, 46.

9. Theodore Roszak, *The Making of a Counter Culture: Reflections on the Technocratic Society and Its Youthful Opposition* (Garden City, NY: Doubleday, 1969), 4–5.

10. Barbara W. Wyden, "A Child on Drugs," *New York Times*, August 20, 1967, SM63.

11. Ibid.

12. "Report Says More Youths Sniff Glue," *Washington Post*, October 3, 1967, B2.

13. Douglas Robinson, "New Westport Arrests Expected in Teen-Age Narcotics Scandal," *New York Times*, November 3, 1967, 44. See also "Report Says More Youths Sniff Glue," *Washington Post*, October 3, 1967, B2; John Kifner, "City Worker Hides Eleven Young Fugitives," *New York Times*, September 1, 1967, 1; John Kifner, "Seven of Twelve Youth House Fugitives Returned to Court," *New York Times*, September 2, 1967, 20.

14. Howard A. Rusk, "Interfaith Neighbors," *New York Times*, November 19, 1967, 83.

15. Herbert Swope, "Deadly Gadgets," *New York Times*, January 30, 1968, 40.

16. Ibid.; "Glue-Sniffer Found Dead," *New York Times*, May 24, 1968, 23. See also Richard Corrigan, "Bid to Retain 6 Percent Interest in Virginia Fails," *Washington Post*, February 9, 1968, B4; Jack Eison, "Health Bill Praised, Financing Rapped," *Washington Post*, March 7, 1968, B2; "Action on Statewide Bills by Virginia Assembly Listed," *Washington Post*, March 10, 1968, C11; "Glue-Sniffing Is Hinted in Va. Death," *Washington Post*, May 24, 1968, B5; Helen Dewar, "New Laws Take Effect in Virginia Starting June 28," *Washington Post*, June 20, 1968, F1, G2.

17. David K. Shipler, "Bronx Glue Angels Fly High," *New York Times*, August 3, 1968, 27.

18. Ibid. See also Irving Spiegel, "Fifty Park Slope Youths in Battle; Two Are Arrested, Two Hurt," *New York Times*, August 11, 1968, 42; M. S. Handler, "Women Give View on Sterilization," *New York Times*, September 1, 1968, 25; Meredith Platt, "Does Your Child Take Dope?" *Washington Post*, June 13, 1968, F1.

19. Musto, *American Disease*, 244–53; King County Bar Association, "Drugs and the Drug Laws," 27–28. See Peniel E. Joseph, *Waiting 'til the Midnight Hour: A Narrative History of Black Power in America* (New York: Henry Holt, 2005); Michael K. Honey, *Going down Jericho Road: The Memphis Strike, Martin Luther King's Last Campaign* (New York: W. W. Norton, 2007).

20. *People v. Orozco*, 266 Cal App 2d 507 (1968); "People v. Orozco (1968)," *American Law Reports* 3d, vol. 32 (Rochester, NY: Lawyers Co-operative Publishing, 1970), 1429.

21. "People v. Orozco (1968)," *West's California Reporter*, vol. 72 (St. Paul, MN: West Publishing Co., 1969), 453.

22. Ibid., 455.

23. Ibid., 456.

24. "People v. Orozco (1968)," *American Law Reports* 3d, 32:1429–37; "People v. Orozco (1968)," *West's California Reporter*, vol. 72 (St. Paul, MN: West Publishing Co., 1969), 452–57; J. A. Bryant, "Penal Offense of Sniffing Glue or Similar Volatile Intoxicants," *American Law Reports* 3d ser., vol. 32 (Rochester, NY: Lawyers Co-operative Publishing, 1970), 1438–40.

25. "Glue Sniffing Causes Chromosome Damage, Medical Study Finds," *Wall Street Journal*, February 25, 1969, 5; Nate Haseltine, "Glue Sniffing May Damage Chromosome," *Washington Post*, February 27, 1969, G6; William Greider, "Wounded Knee Still Festers," *Washington Post*, February 23, 1969, 25; Ann Landers, "If You Observe These Symptoms, Watch for Drugs," *Washington Post*, November 19, 1968, B4.

26. "Study Finds Protests Widespread," *Washington Post*, March 3, 1969, A6.

27. Helen Dewar, "Godwin Praises Assembly's Constitution Efforts," *Washington Post*, April 19, 1969, 26.

28. Fort, *Pleasure Seekers*, 209. Much of this analysis takes its cue from a fifteen-part investigative series on "Children in Trouble," by the *Christian Science Monitor*, the scope of which moves far afield of glue sniffing to tackle the broader problem of how youth—and juvenile delinquents in particular—are treated. Three of the fifteen parts were particularly helpful: Howard Jones, "What about Reform Schools?" *Christian Science Monitor*, May 5, 1969, 9; Howard Jones, "Who Are the Real Delinquents?" *Christian Science Monitor*, May 19, 1969, 9; Howard Jones, "Efforts to Close the Police-Teen Gap," *Christian Science Monitor*, June 16, 1969, 9.

29. Myra MacPherson, "Pot-Smoking Kids: Why Do They Do It?" *Washington Post*, July 9, 1969, D1.

30. Herman W. Land, *What You Can Do about Drugs and Your Child* (New York: Hart Publishing, 1969), 10–11, 17, 63–68. Brill didn't fare much better at the Rutgers Symposium on Drug Abuse that year. He described the long, virtually continuous history of drug abuse and then concluded, somehow, that "there is nothing in history to support the often-quoted view that an outbreak of drug abuse is an irreversible process." He used as an example reports that "Communist China has succeeded in eradicating even opium smoking." Significantly, however, Brill only mentioned glue sniffing in passing, as did several others at the conference; no one made it the priority as they would have done earlier in the 1960s. See Henry Brill, "Recurrent Patterns in the History of Drugs of Dependence and Some Interpretations," in *Drugs and Youth: Proceedings of the Rutgers Symposium on Drug Abuse*, ed. J. R. Wittenborn, Henry Brill, Jean Paul Smith, and Sarah A. Wittenborn (Springfield, IL: Charles C. Thomas, 1969), 8–26. Quotes from p. 24.

31. "Drug Agency Links Spray Can Sniffing to Forty-Two Youth Deaths," *New York Times*, May 10, 1969, 62.

32. Edwin Derby, "Glue Sniffers Beware," *Washington Post*, September 14, 1969, 153.

33. Ibid.

34. Ibid.; John B. Anderson, "Glue-Sniffing Can Be Eliminated," House of Representatives, Extensions of Remarks, *Congressional Record*, July 31, 1969, 21735–36; Derby, "Glue Sniffers Beware," 153.

35. Joshua Lederberg, "A Good Try Poses Problem," *Washington Post*, October 11, 1969, A23.

36. "Public Law 91-113, November 6, 1969," *United States Statutes at Large, 1969*, vol. 83 (Washington, DC: USGPO, 1970), 187-90.

37. "Danville Names Negro Planner," *Washington Post*, October 16, 1969, B2; "Fewer than Forty Gather to Discuss Drug Abuse in Arlington Schools," *Washington Post*, November 18, 1969, C2; Jack H. Morris, "Panicked Parents: Hastily Formed Groups Spring Up across U.S. to Fight Drug Problem," *Wall Street Journal*, December 16, 1969, 1; William Chapman, "Hill Told of Pot Smoking by Mylai GIs," *Washington Post*, March 25, 1970, A1; George W. Ashworth, "What Is Being Done about It," *Christian Science Monitor*, July 22, 1970, 7; "Drug Report Hits Abstinence Stress," *Washington Post*, September 6, 1970, 4; Selig S. Harrison, "Stiff Laws, Tradition Win Japan Drug War," *Washington Post*, November 29, 1970, 1.

## Notes to Chapter 10

1. "Glue Sniffer Linked to Blood Block," *Washington Post*, November 22, 1970, 3.
2. Ibid.
3. Ted Rubin and John Babbs, "The Glue Sniffer," *Federal Probation* 34 (1970): 25.
4. Ibid., 23-28. See also chapter 9, note 28 for more on the *Christian Science Monitor*'s investigative reporting. The final version of the Denver glue-sniffing report appeared in 1968. See Ted Rubin, *Glue Sniffers: A Social-Psychological Assessment of Alternative Juvenile Court Rehabilitation Approaches* (Denver: Denver Juvenile Court, 1968).
5. Rubin and Babbs, "The Glue Sniffer," 25-26.
6. Ibid., 25-26.
7. Ibid., 26.
8. "Biographical Note," Patsy Mink Papers, MS 105, Sophia Smith Collection, Smith College Special Collections, Northampton, Massachusetts.
9. Ibid.
10. "HR 12751," House of Representatives, Public Bills and Resolutions, *Congressional Record*, July 10, 1969, 19143-44; Patsy Mink, "A Bill to Ban Glue and Paint Products Containing Toxic Solvents," House of Representatives, *Congressional Record*, July 10, 1969, 19078-80; Tomi Knaefler, "Isle Youngsters Turn to Drugs in Growing Numbers," *Honolulu Star-Bulletin*, June 23, 1969, 1.
11. "Testimony by Representative Patsy T. Mink before the Subcommittee on Public Health and Welfare of the House Committee on Interstate and Foreign Commerce on Legislation to Protect Children from Glue-Sniffing," February 17, 1970, Patsy Mink Papers, MS 105, Sophia Smith Collection, Special Collections, Smith College, Northampton, Massachusetts.
12. Ibid.
13. Patsy Mink, "Protect Children from Glue Sniffing," House of Representatives, *Congressional Record*, March 11, 1970, 6842-43.
14. "Letter from Joshua Lederberg to The Hon. John Harman, Chairman, Subcommittee on Public Health and Welfare," Patsy Mink Papers, MS 105, Sophia Smith Collection, Special Collections, Smith College, Northampton, Massachusetts.
15. Ibid.

16. *James Joseph Ward v. Maryland*, 9 Md. App. 583 (1970).
17. Ibid.
18. "Public Law 91-601, December 30, 1970," *United States Statutes at Large, 1970,* vol. 84, pt. 2 (Washington, DC: USGPO, 1971), 1670-74; "Public Law 100-695, November 18, 1988," *United States Statutes at Large, 1988*, vol. 105, pt. 5 (Washington, DC: USGPO, 1990), 4568-70; "Public Law 103-267, June 16, 1994," *United States Statutes at Large, 1994*, vol. 108, pt. 1 (Washington, DC: USGPO, 1995), 722-29. Congress would continue this trend in the decades to come. In 1988, legislators again amended the Federal Hazardous Substances Act "to require the labeling of chronically hazardous art materials," but again the law was directed at long-term exposure during normal use and accidental ingestion. In 1994, Congress again amended the act, this time with the Child Safety Protection Act, requiring potentially hazardous toys and games to be labeled properly. Again, though, the law dealt with choking hazards, balls, and bicycle helmets. It dealt with the susceptibility of children to accidental problems that came in the normal use of certain toys and games. Intentional misuse was once again never addressed.
19. "Youth and Marijuana," *Washington Post*, September 15, 1971, E15.
20. Ann Landers, "Crests," *Washington Post*, July 6, 1972, C10.
21. LaBarbara Bowman, "'Saddest' Cases Are Heard in Juvenile Court," *Washington Post*, September 28, 1972, F1, F2; Eleanor Gurewitsch, "Families Aid Children in Trouble," *Christian Science Monitor*, April 25, 1973, 8.
22. *Marilyn Harless v. Boyle-Midway Division, American Home Products*, 594 F.2d 1051 (1979).
23. Ibid.
24. Ibid.
25. Marc Lalonde, *Final Report of the Commission of Inquiry into the Non-medical Use of Drugs*, December 14, 1973 (Ottawa: Commission of Inquiry into the Non-medical Use of Drugs, 1973), Appendix B8.
26. King County Bar Association, "Drugs and the Drug Laws," 29.
27. Musto, *American Disease*, 251-65; Spillane, "Federal Policy in the Post-Anslinger Era," 209-216; King County Bar Association, "Drugs and the Drug Laws," 29-30. See also Steven B. Duke and Albert C. Gross, *America's Longest War: Rethinking Our Tragic Crusade against Drugs* (New York: Tarcher Books, 1994); William Weir, *In the Shadow of the Dope Fiend: America's War on Drugs* (North Haven, CT: Archon Books, 1995).
28. King County Bar Association, "Drugs and the Drug Laws," 31.
29. For more on drug laws, race, and the creation of a permanent underclass, see Alexander, *The New Jim Crow*.
30. Jimmie L. Reeves and Richard Campbell, *Cracked Coverage: Television News, the Anti-cocaine Crusade, and the Reagan Legacy* (Durham: Duke University Press, 1994), 15-47, 73-103; Musto, *American Disease*, 265-77; Spillane, "Federal Policy in the Post-Anslinger Era," 216-20; King County Bar Association, "Drugs and the Drug Laws," 30-31.
31. Janis Johnson, "Churches in Area Locking Doors to Combat Crime and Vandalism," *Washington Post*, November 29, 1974, A1; Harold J. Logan, "Maryland Deaths from Freon Rise," *Washington Post*, December 22, 1974, C2; Douglas C. Lyons, "Fairfax Girl, Thirteen, Found Dead," *Washington Post*, March 13, 1975, C1, C8; Thomas Grubisich, "Life of Crime Began at Nine," *Washington Post*, August 24, 1976, A1.
32. Sharp and Brehm, *Review of Inhalants*, 1-227 (21). For another example of later academic work on the subject, see Denis J. O'Connor, *Glue Sniffing and Volatile Substance*

*Abuse by Schoolchildren and Adolescents* (PhD diss., University of Newcastle-upon-Tyne, England, 1986).

33. "Now I Wanna Sniff Some Glue," The Ramones, *The Ramones*, 1976.

## Notes to Conclusion

1. For more on the influenza epidemic, see John M. Barry, *The Great Influenza: The Epic Story of the Greatest Plague in History* (New York: Viking, 2004); Alfred W. Crosby, *America's Forgotten Pandemic: The Influenza of 1918* (New York: Cambridge University Press, 2003).

2. Kenneth Thompson, *Moral Panics* (New York: Routledge, 1998), 43.

3. Alexander, *The New Jim Crow*; Craig Reinarman and Harry G. Levine, *Crack in America: Demon Drugs and Social Justice* (Berkeley: University of California Press, 1997), 1–52. Chomsky has discussed this deficit in several venues but perhaps most succinctly in conversation with Daniel Mermet, "Democracy's Invisible Line," *Le Monde Diplomatique* (August 2007), 1, accessed July 29, 2013, http://mondediplo.com/2007/08/02democracy (English translation). See also Goode and Ben-Yehuda, "Moral Panics," 150, 168–69.

4. For more on the development of drug panics from the 1970s forward, from amphetamine, to PCP, to cocaine, and beyond, see Philip Jenkins, *Synthetic Panics: The Symbolic Politics of Designer Drugs* (New York: New York University Press, 1999). For more on the rave culture that developed following the 1980s (in this case as it developed in Canada), see Sean P. Hier, "Raves, Risks and the Ecstacy Panic: A Case Study in the Subversive Nature of Moral Regulation," *Canadian Journal of Sociology* 27 (Winter 2002): 33–57.

## Notes to Appendix 1

1. These descriptions come directly from the National Institute on Drug Abuse monograph. Ellipses in the text omit references to tables or other elements specific to the NIDA document.

2. Erick G. Comstock and Betsy S. Comstock, "Medical Evaluation of Inhalant Abusers," in Sharp and Brehm, *Review of Inhalants*, 56.

3. Ibid.

4. Ibid.

5. Ibid., 66.

6. James V. Bruckner and Richard G. Peterson, "Review of the Aliphatic and Aromatic Hydrocarbons," in ibid., 124.

7. Ibid., 133.

8. Ibid., 136.

9. Comstock and Comstock, "Medical Evaluation," ibid., 66.

10. Bruckner and Peterson, "Aliphatic and Aromatic Hydrocarbons," ibid., 138.

# Bibliography

## Newspapers

*Atlanta Journal*
*Atlanta Journal and Constitution*
*Baton Rouge Advocate*
*Birmingham News*
*Christian Science Monitor*
*Denver Post*
*Honolulu Star-Bulletin*
*Louisiana Weekly*
*Monroe Morning World*
*Montgomery Advertiser*
*New York Times*
*Rockford Register Star*
*Rome News-Tribune*
*Spokesman-Review*
*Wall Street Journal*
*Washington Post*

## Periodical Press

"Fads: The New Kick." *Time*, February 16, 1962, 55.
"Glue Sniffing: An Adolescent Craze That Is Not Amusing." *Consumer Reports* 28 (January 1963): 40.
"Going to Seed." *Newsweek*, July 22, 1963, 59.
"The New Addicts." *Newsweek*, August 13, 1962, 42.
"Whiff of Innocence." *Time*, August 25, 1967, 51.

## Other Media

*Airplane!* Directed by Jim Abrahams, David Zucker, and Jerry Zucker. Paramount Pictures, 1980.
Bruce, Lenny. "Airplane Glue." *American*, Fantasy Records, 7011, F-2081, 1961.
Fletcher, Grace Nies. *What's Right with Our Young People*. New York: William Morrow, 1966.
Gordon, Dorothy. *Who Has the Answer?* New York: E. P. Dutton, 1965.
Hersey, John. *Too Far to Walk*. New York: Alfred A. Knopf, 1966.

Maugham, Somerset. *On a Chinese Screen*. New York: Doran and Co., 1922.
Pike, James A. *Teen-Agers and Sex*. Englewood Cliffs, NJ: Prentice Hall, 1965.
Purtell, Thelma C. *Tonight Is Too Late*. New York: P. S. Eriksson, 1965.
The Ramones. "Now I Wanna Sniff Some Glue." *The Ramones*, 1976.
Shakespeare, William. *The Winter's Tale*.
Wilde, Oscar. *The Picture of Dorian Gray*. London: Ward, Lock, and Co., 1891.

### Archival Material

H. Ted Rubin Papers. WH2106. Western History Collection. Denver Public Library, Denver, Colorado.
Patsy Mink Papers. MS 105. Sophia Smith Collection. Smith College Special Collections. Northampton, Massachusetts.

### Selected Government Documents

Council of Economic Advisers. *Economic Report of the President, Transmitted to the Congress February 1992*. Washington, DC: US Government Printing Office, 1992 (hereafter USGPO).
Federal Bureau of Investigation. *Crime in the United States: Uniform Crime Reports, 1991*. Washington, DC: Uniform Crime Reporting Program, 1992.
"Project MKULTRA. The CIA's Program of Research into Behavioral Modification. Joint Hearing before the Select Committee on Intelligence and the Subcommittee on Health and Scientific Research of the Committee on Human Resources." US Senate, Ninety-Fifth Congress, First Session, August 8, 1977. Washington, DC: USGPO, 1977.
"Single Convention on Narcotic Drugs, 1961." United Nations. Accessed July 26, 2013. http://www.incb.org/documents/Narcotic-Drugs/1961-Convention/convention_1961_en.pdf.
US Bureau of the Census. *1970 Census of Population*. Vol. 1, *Characteristics of the Population*. Part A, *Number of Inhabitants*. Section 1, *United States, Alabama-Mississippi*. Washington, DC: USGPO, 1972.
———. *1970 Census of Population, Subject Reports: Final Report PC(2)-9A: Low-Income Population*. Washington, DC: USGPO, 1973.
———. *1970 Census of Population, Subject Reports: Final Report PC(2)-9B: Low-Income Areas in Large Cities*. Washington, DC: USGPO, 1973.
US Department of Justice. *Sourcebook of Criminal Statistics, 1990*. Rockville, MD: Justice Statistics Clearinghouse, 1991.

### Academic Research

Ackerly, William C., and Guadalupe Gibson. "Lighter Fluid 'Sniffing.'" *American Journal of Psychiatry* 120 (May 1964): 1056–61.
Aichhorn, August. *Wayward Youth*. New York: Viking Press, 1935.

Alex, Ted. "Denver Juvenile Court Glue-Sniffing Project." In *Presentations of a Conference on Inhalation of Glue Fumes and Other Substance Abuse Practices among Adolescents*, ed. Ted Rubin, 96–103. Denver: Denver Juvenile Court, 1967.
Anderson, Per, and Birger R. Kaada. "The Electroencephalogram in Poisoning by Lacquer Thinner." *Acta Pharmacologica et Toxicologica* 9 (April 1953): 125–30. Andre, McKenzie, Kashef Ijaz, Jon D. Tillinghast, Valdis E. Krebs, Lois A. Diem, Beverly Metchock, Theresa Crisp, and Peter D. McElroy. "Transmission Network Analysis to Complement Routine Tuberculosis Contact Investigations." *American Journal of Public Health* 97 (March 2007): 470–77.
Barker, Gordon H., and W. Thomas Adams. "Glue Sniffers." *Sociology and Social Research* 47 (April 1963): 298–310.
Barman, Martin L., and Donn B. Beedle. "Acute and Chronic Effects of Glue Sniffing." *California Medicine* 100 (January 1964): 19–22.
Barnes, Gordon E., and Brent A. Vulcano. "Bibliography of the Solvent Abuse Literature." *International Journal of the Addictions* 14, no. 3 (1979): 401–21.
Bartlett, Sylvan, and Fernando Tapia. "Glue and Gasoline 'Sniffing,' The Addiction of Youth." *Missouri Medicine* 63 (April 1966): 270–72.
Bernabeu, Ednita P. "Underlying Ego Mechanisms in Delinquency." *Psychoanalytic Quarterly* 27 (July 1958): 383–96.
Bernard, Leopold. *Les odeurs dans les romans de Zola*. Montpellier: C. Coulet, 1889.
Bleuler, Eugen. *Dementia Praecox, or the Group of Schizophrenias*. Translated by J. Zinkin. New York: International Universities Press, 1950.
———. *Lehrbuch de Psychiatrie*. Berlin: Springer-Verlag, 1916.
Brecher, Edward M., and the Editors of *Consumer Reports Magazine*. *The Consumers Union Report: Licit and Illicit Drugs*. See esp. pt. 6, "Inhalants, Solvents and Glue-sniffing," incl. ch. 43, "The historical antecedents of glue-sniffing," and ch. 44, "How to Launch a Nationwide Drug Menace." http://www.druglibrary.org/schaffer/Library/studies/cu/cumenu.htm.
Brill, A. A. "The Sense of Smell in the Neuroses and Psychoses." *Psychoanalytic Quarterly* 1 (April 1932): 7–42.
Brill, Henry. "Recurrent Patterns in the History of Drugs of Dependence and Some Interpretations." In *Drugs and Youth: Proceedings of the Rutgers Symposium on Drug Abuse*, edited by J. R. Wittenborn, Henry Brill, Jean Paul Smith, and Sarah A. Wittenborn, 8–26. Springfield, IL: Charles C. Thomas, 1969.
Brotman, Richard. "Adolescent Substance Use: A Growing Form of Dissent." In Rubin, *Presentations of a Conference*, 37–43.
Chapel, James L., and Daniel W. Taylor. "Glue Sniffing." *Missouri Medicine* 65 (April 1965): 288–92.
Chein, Isidor, and Eva Rosenfeld. "Juvenile Narcotics Use." *Law and Contemporary Problems* 22 (Winter 1957): 52–68.
Christiansson, G., and B. Karlsson. "Sniffing, Method of Intoxication among Children." *Svensk Lakartidn* 54 (January 1957): 33–44.
Clinger, Orris W., and Nelson A. Johnson. "Purposeful Inhalation of Gasoline Vapors." *Psychiatric Quarterly* 25 (October 1951): 557–67.
Coodley, Alfred E. "Current Aspects of Delinquency and Addictions." *Archives of General Psychiatry* 4 (June 1961): 632–40.

Corliss, Leland M. "A Review of the Evidence on Glue-Sniffing—A Persistent Problem." *Journal of School Health* 35 (December 1965): 442–49.

Davis, Richard. "Report on the Problem of Glue Sniffing in Children and the Work of the New York City Police Department and Its Youth Investigation Bureau in Combating This Problem." In Rubin, *Presentations of a Conference*, 104–10.

Deisher, Robert W., Albert J. Schroeder, V. Robert Allen, Harry Bakwin, Victor Eisner, Dale C. Garell, S. L. Hammar, Sprague W. Hazard, Thomas E. Shaffer, John Allen Welty, Charles Louis Wood, Charles Keck, and Graham Blaine. "Drug Abuse in Adolescence: The Use of Harmful Drugs, a Pediatric Concern." *Pediatrics* 44 (July 1969): 131–41.

Dodds, Josiah, and Sebastiano Santostefano. "A Comparison of the Cognitive Functioning of Glue Sniffers and Non-sniffers." *Journal of Pediatrics* 64 (April 1964): 565–70.

Done, Alan K. "Presentation before Conference on Substance Abuse Practices among Adolescents." In Rubin, *Presentations of a Conference*, 44–50.

Dreisbach, Robert H., ed. *Handbook of Poisoning: Diagnosis and Treatment*. Los Altos, CA: Lange Medical Publications, 1961.

"Drugs and Substance Abuse." *Florida Health Notes* 61 (October 1969): 272.

Easson, William M. "Gasoline Addiction in Children." *Pediatrics* 29 (February 1962): 250–54.

Edwards, Robert V. "A Case Report of Gasoline Sniffing." *American Journal of Psychiatry* 117 (December 1960): 555–57.

Ellis, Havelock. *Sexual Selection in Man*. New York: F. A. Davis, 1906.

Ellison, Willie S. "Portrait of a Glue Sniffer." *Crime and Delinquency* 11, no. 4 (1965): 394–99.

Ely, Dale F. "Substance Abuse—A Community Problem: The School." In Rubin, *Presentations of a Conference*, 124–25.

Engel, George L., and John Romano. "Delirium: A Syndrome of Cerebral Insufficiency." *Journal of Chronic Disorders* 9 (March 1959): 260–77.

Expert Committee on Drugs Liable to Produce Addiction. *Expert Committee on Drugs Liable to Produce Addiction: Second Report*. World Health Organization Technical Report Series No. 21. Geneva: World Health Organization, 1950.

Faucett, Robert L., and Reynold A. Jensen. "Psychologic Aspects of Pediatrics: Addiction to the Inhalation of Gasoline Fumes in a Child." *Journal of Pediatrics* 41 (September 1952): 364–68.

Ferenczi, Sándor. *First Contributions to Psycho-analysis*. Translated by Ernest Jones. 1916. London: Karnac Books, 1994.

Fliess, Wilhelm. *Die Beziehungen zwischen Nase und weiblichen Geschlechtsorganen, In ihrer biologischen Bedeutung dargestellt*. Leipzig: Deuticke, 1897.

Freedman, Alfred M., and Ethel A. Wilson. "Childhood and Adolescent Addictive Disorders." *Pediatrics* 34 (1964): 283–92.

Freud, Sigmund. "Bemerkungen über einen Fall von Zwangsneurose." *Gesammelte Schriften* 8 (1908): 350.

———. *Three Contributions to the Theory of Sex*. Translated by A. A. Brill. New York: Nervous and Mental Disease Publishing Co., 1920.

Friedman, Paul. "Some Observations on the Sense of Smell." *Psychoanalytic Quarterly* 28 (July 1959): 307–29.

Gellman, Vera. "Glue-Sniffing among Winnipeg School Children." *Canadian Medical Association Journal* 98 (February 24, 1968): 411-13.
Gerard, Donald, and Conan Kornetsky. "Adolescent Opiate Addiction." *Psychiatric Quarterly* 29 (July 1955): 457-86.
Gioscia, Victor. "Glue Sniffing: Exploratory Hypotheses on the Psychosocial Dynamics of Respiratory Intrajection." In Rubin, *Presentations of a Conference*, 60-73.
Glaser, Frederick. "Inhalation Psychosis and Related States." *Archives of General Psychiatry* 14 (March 1966): 315-22.
Glaser, Helen H., and Oliver N. Massengale. "Glue-Sniffing in Children: Deliberate Inhalation of Vaporized Plastic Cements." *Journal of the American Medical Association* 181 (July 28, 1962): 300-303.
Glatt, M. M. "Drug Treatment Centres." *British Medical Journal* 3 (July 22, 1967): 242.
Gleason, Marion N., Robert E. Gosselin, and Harold Carpenter Hodge. *Clinical Toxicology of Commercial Products*. Baltimore: Williams and Wilkins, 1957.
"Glue-Sniffing." *Journal of the American Medical Association* 181 (July 28, 1962): 333.
"Glue Sniffing ... The Quest for Ecstacy." *Georgia's Health* 47 (February 1967): 1-2.
Grabski, D. A. "Toluene Sniffing Producing Cerebellar Degeneration." *American Journal of Psychiatry* 118 (November 1961): 461-62.
Grant, W. B. "Inhalation of Gasoline Fumes by a Child." *Psychiatric Quarterly* 36 (July 1962): 555-57.
Green, Robert C. Jr. "Nutmeg Poisoning." *Journal of the American Medical Association* 171 (November 7, 1959): 1342-44.
Greenburg, Leonard, May R. Mayers, Harry Heimann, and Samuel Moskowitz. "The Effects of Exposure to Toluene in Industry." *Journal of the American Medical Association* 118 (February 12, 1942): 573-78.
Hambridge, Gare. "Scents That Make Dollars." *World's Work* 62 (August 1931): 274.
Hanson, Robert C. "Explaining Glue Sniffing and Related Juvenile Delinquency." In Rubin, *Presentations of a Conference*, 82-95.
Hinton, Pamela, and Bernd Koch. "Hemolytic Uremic Syndrome: Report of Two Cases." *Canadian Medical Association Journal* 98 (April 27, 1968): 819-26.
Houseworth, Donald E. "A Study of Retreatism in Glue Sniffing and Non-Glue-Sniffing Delinquents in Utah." PhD diss. Brigham Young University, 1968.
Jacobziner, H., and H. W. Raybin. "Glue-Sniffing." *New York State Journal of Medicine* 62 (October 15, 1962): 3294-96.
Keeler, Martin H. "The Use of Hyoscyamine as a Hallucinogen and Intoxicant." *American Journal of Psychiatry* 124 (December 6, 1967): 852-54.
Keeler, Martin H., and Clifford B. Reifler. "The Occurrence of Glue Sniffing on a University Campus." *Journal of the American College Health Association* 16 (October 1967): 69-70.
Koch-Grünberg, Theodor. *From Roraima to the Orinoco*. New York: Cambridge University Press, 2009.
Kupperstein, Lenore. "Summary of Discussion from Workshop 'B.'" In Rubin, *Presentations of a Conference*, 132-33.
Kupperstein, Lenore R., and Ralph M. Susman. "A Bibliography on the Inhalation of Glue Fumes and Other Toxic Vapors—A Substance Abuse Practice among Adolescents." *International Journal of the Addictions* 3 (Spring 1968): 177-97.

Lalonde, Marc. *Final Report of the Commission of Inquiry into the Non-medical Use of Drugs. December 14, 1973*. Ottawa: Commission of Inquiry into the Non-medical Use of Drugs, 1973.

Latta, S. J., and L. Davies. "Effects on the Blood and Hemopoietic Organs of the Albino Rat of Repeated Administration of Benzene." *Archives of Pathology and Laboratory Medicine* 31 (1941): 55–67.

Lawton, James J., and Carl P. Malmquist. "Gasoline Addiction in Children." *Psychiatric Quarterly* 35 (July 1961): 555–61.

Lindesmith, Alfred R. *Addiction and Opiates*. 1947. Chicago: Aldine Publishing, 1968.

———. "A Sociological Theory of Drug Addiction." *American Journal of Sociology* 43 (January 1938): 593–613.

Lockhead, H. B., and H. P. Close. "Ethylene, Dichloride Plastic Cement: A Case of Fatal Poisoning." *Journal of the American Medical Association* 146 (August 4, 1951): 1323.

Lohman, Joseph D. "Youth: New Problems and New Directions." In Rubin, *Presentations of a Conference*, 19–36.

Lowry, James V. "Hospital Treatment of the Narcotic Addict." *Federal Probation* 20 (December 1956): 42–51.

Luger, Milton. "Substance Abuse—A Community Problem." In Rubin, *Presentations of a Conference*, 126–28.

Machle, Willard. "Gasoline Intoxication." *Journal of the American Medical Association* 117 (December 6, 1941): 1965–72.

Meloff, William A. "Deviant Attitudes and Behavior of Glue-Sniffers in Comparison with Similar and Different Class Peer Groups." In Rubin, *Presentations of a Conference*, 74–81.

Merry, Julius. "Glue Sniffing and Heroin Abuse." *British Medical Journal* 2 (May 6, 1967): 360.

Merry, Julius, and Nicholas Zachariadis. "Addiction to Glue Sniffing." *British Medical Journal* 2 (December 1, 1962): 1448.

*1974 Summary of Florida State Plan for Drug Abuse Prevention*. Tallahassee: Florida Department of Health and Rehabilitative Services, 1974. Florida State Library.

Nitsche, C. J., and J. F. Robinson. "A Case of Gasoline Addiction." *American Journal of Orthopsychiatry* 24 (April 1959): 417–19.

Norman, Connolly. "Variations in Form of Mental Affections in Relation to the Classification of Insanity." *Dublin Journal of Medical Science* 83 (1887): 228–35.

Nunn, John A., and Frank M. Martin, "Gasoline and Kerosene Poisoning in Children." *Journal of the American Medical Association* 103 (August 18, 1934): 472–74.

O'Brien, Frederick. *Atolls of the Sun*. New York: Century Co., 1922.

von Oettingen, Wolfgang F., Paul A. Neal, and Dennis D. Donahue. "The Toxicity and Potential Dangers of Toluene." *Journal of the American Medical Association* 118 (February 12, 1942): 579–84.

Oldschool, Oliver. "Ether Sniffers." *Journal of the American Medical Association* 197 (July 11, 1966): 13. Originally published in *Esquire*, 1824.

Pescor, Michael J. *A Statistical Analysis of the Clinical Records of Hospitalized Drug Addicts*. Supplement No. 143. Public Health Reports. Washington, DC: USGPO, 1943.

Pierson, Howard W. "Glue Sniffing: A Hazardous Hobby." *Journal of School Health* 34 (May 1964): 252.

Powars, Darleen. "Aplastic Anemia Secondary to Glue Sniffing." *New England Journal of Medicine* 273 (September 23, 1965): 700–702.
Press, Edward. "Glue Sniffing." *Journal of Pediatrics* 63 (September 1963): 516–18.
Press, Edward, and Alan K. Done. "Solvent Sniffing: Physiologic Effects and Community Control Measures for Intoxication from the Intentional Inhalation of Organic Solvents, I." *Pediatrics* 39 (March 1967): 451–61.
———. "Solvent Sniffing: Physiologic Effects and Community Control Measures for Intoxication from the Intentional Inhalation of Organic Solvents, II." *Pediatrics* 39 (April 1967): 611–22.
Pruitt, Maurice. "Bizarre Intoxications." *Journal of the American Medical Association* 171 (December 26, 1959): 2355.
Radó, Sándor. "The Psychoanalysis of Pharmacothymia (Drug Addiction)." *Psychoanalytic Quarterly* 2, no. 1 (1933): 1–23.
Ravin, Louis J., and Peter D. Bernardo. "Pharmaceutical Sciences—1967: A Literature Review of Pharmaceutics." *Journal of Pharmaceutical Sciences* 57 (July 1968): 1075–97.
Rawlin, John William. "Identification of the Problem of Substance Abusing Adolescents: The Problem of Amphetamine Abuse." In Rubin, *Presentations of a Conference*, 119–23.
Redl, Fritz. "The Psychology of Gang Formation and the Treatment of Juvenile Delinquents." In *Psychoanalytic Study of the Child*, edited by Ruth S. Eissler, Albert J. Solnit, Anna Freud, and Marianne Kris, 1:367–77. New York: International Universities Press, 1945.
Romano, John, and George L. Engel. "Delirium: I. Electroencephalographic Data." *Archives of Neurologic Psychiatry* 51 (April 1944): 356–77.
Rorke, Joseph F. "Plastic Cement Fumes." *Journal of the American Medical Association* 173 (July 16, 1960): 1277.
Rubin, Ted. "An Appeal for Early and Intensive Intervention." In Rubin, *Presentations of a Conference*, 139–40.
———. *Glue Sniffers: A Social-Psychological Assessment of Alternative Juvenile Court Rehabilitation Approaches*. Denver: Denver Juvenile Court, 1968.
Rubin, Ted, ed. *Presentations of a Conference on Inhalation of Glue Fumes and Other Substance Abuse Practices among Adolescents*. Denver: Denver Juvenile Court, 1967.
Rubin, Ted, and John Babbs. "The Glue Sniffer." *Federal Probation* 34 (1970): 23–28.
Schmideberg, Melitta. "The Psychoanalysis of Delinquents." *American Journal of Orthopsychiatry* 23 (January 1953): 13–21.
Schneck, J. M. "Chloroform Habituation: With a Case Report of Its Occurrence in a Case of Schizophrenia." *Bulletin of the Menninger Clinic* 9 (January 1945): 12–17.
Shanholtz, Mack I. "Glue Sniffing: A New Symptom of an Old Disease." *Virginia Medical Monthly* 95 (May 1968): 304–5.
Sharp, Charles William, and Mary Lee Brehm. *Review of Inhalants: Euphoria to Dysfunction*. NIDA Research Monograph 15. Department of Health, Education, and Welfare. October 1977.
Silverstein, Harry. "Summary of Discussion from Workshop 'A.'" In Rubin, *Presentations of a Conference*, 129–31.
"'Sniffing' Contents of Aerosol Cans Leads to Death." *Georgia's Health* 47 (October 1967): 3.

Sobolevitch, Robert H. "Juvenile Substance Abuse in Philadelphia." In Rubin, *Presentations of a Conference*, 111–18.
Sokol, Jacob. "Glue Sniffing among Juveniles." *American Journal of Correction* 27 (November-December 1965): 18–21.
Stefansson, Vilhjalmur. *My Life with the Eskimo*. New York: Macmillan, 1919.
Sterling, James W. "A Comparative Examination of Two Modes of Intoxication: An Exploratory Study of Glue Sniffing." *Journal of Criminal Law, Criminology, and Political Science* 55 (March 1964): 94–99.
Tefferteller, Ralph. "Summary of Discussion from Workshop 'C.'" In Rubin, *Presentations of a Conference*, 134–35.
Thomas, Lester G. "The Denver Glue Sniffing Project." *Juvenile Court Judges Journal* 18 (July 1967): 46–49.
Thompson, Kenneth. *Moral Panics*. New York: Routledge, 1998.
Todd, John. "'Sniffing' and Addiction." *British Medical Journal* 4 (October 26, 1968): 255–56.
Tolan, Eliere J., and Fredrich A. Lingel. "'Model Psychosis' Produced by Inhalation of Gasoline Fumes." *American Journal of Psychiatry* 120 (February 1964): 757–61.
Unwin, J. Robertson. "Illicit Drug Use among Canadian Youth: Part I." *Canadian Medical Association Journal* 98 (February 24, 1968): 402–7.
Verhulst, Henry L., and John J. Crotty. "Glue-Sniffing, II." *Bulletin: National Clearinghouse for Poison Control Centers* (July–August 1964): 1–8.
———. "Glue-Sniffing Deterrent." *Bulletin: National Clearinghouse for Poison Control Centers* (November–December 1969): 4–5.
von Krafft-Ebing, Richard. *Psychopathia Sexualis*. 1886. Burbank, CA: Bloat Books, 1999.
Whelton, Michael J. "Glue Sniffing." *British Medical Journal* 2 (November 24, 1962): 1404.
Wilson, Rex H. "Toluene Poisoning." *Journal of the American Medical Association* 123 (December 25, 1943): 1106–8.
Wilson, Roberta. "Summary of Discussion from Workshop 'D.'" In Rubin, *Presentations of a Conference*, 136–38.
Winek, Charles L, Wellon D. Collom, and Cyril H. Wecht. "Fatal Benzene Exposure by Glue-Sniffing." *The Lancet*, March 25, 1967, 683.
———. "Toluene Fatality from Glue Sniffing." *Pennsylvania Medicine* 71 (April 1968): 81.
Wolf, M. A., V. K. Row, D. D. McCollister, R. L. Hollingsworth, and F. Oyen. "Toxicological Studies of Certain Alkylated Benzenes and Benzene." *Archives of Industrial Health* 14 (October 1956): 387–98.

## Other Primary Sources

Chadwick, Bruce A., and Tim B. Heaton. *Statistical Handbook on the American Family*. Phoenix: Oryx Press, 1999.
*Constitution of the State of Georgia*. Atlanta: Secretary of State, 2009.
Department of Education. *Status Report: Drug Education, 1971–72*. Tallahassee: State of Florida, January 1972.
*Drug Abuse in Florida*. Tallahassee: Governor's Task Force on Narcotics, May 1970.
*Florida Juvenile Court Statistics, 1965, 1966, 1967*. Tallahassee: State Department of Public Welfare, 1968. Florida State Library.

Fort, Joel. *The Pleasure Seekers: The Drug Crisis, Youth, and Society*. Indianapolis: Bobbs-Merrill, 1969.
Fortson, Ben W. *Georgia's Official Register, 1967–1968*. Atlanta: Department of Archives and History, 1967.
Harrington, Michael. *The Other America: Poverty in the United States*. 1962. New York: Penguin, 1981.
Joint Committee of the American Bar Association and the American Medical Association on Narcotic Drugs. *Drug Addiction: Crime or Disease? Interim and Final Reports*. Bloomington: University of Indiana Press, 1961.
Keniston, Kenneth. *The Uncommitted: Alienated Youth in American Society*. New York: Harcourt Brace, 1965.
———. *Young Radicals: Notes on Committed Youth*. New York: Harcourt Brace, 1968.
Land, Herman W. *What You Can Do about Drugs and Your Child*. New York: Hart Publishing, 1969.
"The LSD Story." *Dragnet 1967*. January 12, 1967. Internet Movie Database. Accessed August 20, 2013. http://www.imdb.com/title/tt0565680/.
Moody, Scott, ed. *Facts and Figures on Government Finance*. 33rd edition. Washington, DC: Tax Foundation, 1999.
"Port Huron Statement of the Students for a Democratic Society, 1962." Accessed August 17, 2013. http://coursesa.matrix.msu.edu/~hst306/documents/huron.html.
Roszak, Theodore, ed. *The Dissenting Academy*. New York: Pantheon, 1968.
Roszak, Theodore. *The Making of a Counter Culture: Reflections on the Technocratic Society and Its Youthful Opposition*. Garden City, NY: Doubleday, 1969.
*Runningawayness*. Washington, DC: USGPO, 1967.
Smith, Kline and French Laboratories. *Drug Abuse: Escape to Nowhere*. Washington, DC: National Education Association, 1967.
Stanley, Harold W., and Richard G. Niemi. *Vital Statistics on American Politics, 2011–2012*. Thousand Oaks, CA: CQ Press, 2011.
State Department of Public Welfare. *Mississippi Youth Court Statistics*. Jackson: Mississippi Department of Public Welfare.
University of California Academic Senate. *University of California: In Memoriam, 1969*. Berkeley: University of California, 1969.
*What We Can Do about Drug Abuse*. New York: Public Affairs Pamphlets, 1967.
Wolfe, Tom. *The Electric Kool-Aid Acid Test*. 1968. New York: Picador, 2008.

## Secondary Sources

Abrahamson, Mark. "Sudden Wealth, Gratification, and Attainment: Durkheim's Anomie of Influence Reconsidered." *American Sociological Review* 45 (February 1980): 49–57.
Acker, Caroline Jean. *Creating the American Junkie: Addiction Research in the Classic Era of Narcotic Control*. Baltimore: Johns Hopkins University Press, 2002.
———. "Portrait of an Addicted Family: Dynamics of Opiate Addiction in the Early Twentieth Century." In *Altering American Consciousness: The History of Alcohol and Drug Use in the United States, 1800–2000*, edited by Sarah W. Tracy and Caroline Jean Acker, 165–81. Amherst: University of Massachusetts Press, 2004.

Aitken, Stuart. *Geographies of Young People: The Morally Contested Spaces of Identity.* New York: Routledge, 2001.

Alexander, Michelle. *The New Jim Crow: Mass Incarceration in the Age of Colorblindness.* New York: New Press, 2010.

Auerhahn, Kathleen. "The Split Labor Market and the Origins of Antidrug Legislation in the United States." *Law and Social Inquiry* 24 (Spring 1999): 411–40.

Avdela, Efi. "'Corrupting and Uncontrollable Activities': Moral Panic about Youth in Post-Civil War Greece." *Journal of Contemporary History* 43 (January 2008): 25–44.

Barry, John M. *The Great Influenza: The Epic Story of the Greatest Plague in History.* New York: Viking, 2004.

Becker, Howard S. *Outsiders: Studies in the Sociology of Deviance.* New York: Free Press, 1963.

Ben-Yehuda, Nachman. *The Politics and Morality of Deviance: Moral Panics, Drug Abuse, Deviant Science, and Reversed Stigmatization.* Albany: SUNY Press, 1990.

——. "The Sociology of Moral Panics: Toward a New Synthesis." *Sociological Quarterly* 27 (Winter 1986): 495–513.

Best, Joel. *Random Violence: How We Talk about New Crimes and New Victims.* Berkeley: University of California Press, 1999.

Bruce, Lenny. *How to Talk Dirty and Influence People.* New York: Playboy Publishing, 1963.

Campbell, Nancy D. *Discovering Addiction: The Science and Politics of Substance Abuse Research.* Ann Arbor: University of Michigan Press, 2007.

Caputo, Rosario Anthony. *Hillbilly Glue Sniffing and Delinquency: The Social Structuring of Deviant Behavior among Southern White Migrants.* EdD diss. Columbia University Teachers College, 1988.

Carroll, Rebecca. "Under the Influence: Harry Anslinger's Role in America's Drug Policy." In *Federal Drug Control: The Evolution of Policy and Practice,* edited by Jonathon Erlen and Joseph F. Spillane, 61–100. Binghamton, NY: Haworth Press, 2004.

Chomsky, Noam, and Daniel Mermet. "Democracy's Invisible Line." *Le Monde Diplomatique* 1 (August 2007). Accessed July 29, 2013. http://mondediplo.com/2007/08/02democracy (English translation).

Cohen, Stanley. *Folk Devils and Moral Panics: The Creation of the Mods and Rockers.* 1972. New York: St. Martin's Press, 1980.

Collins, Ronald K. L., and David M. Skover. *The Trials of Lenny Bruce: The Fall and Rise of an American Icon.* New York: Sourcebooks, 2002.

Cooper, John Milton. *Woodrow Wilson: A Biography.* New York: Alfred A. Knopf, 2009.

Courtwright, David T. *Dark Paradise: A History of Opiate Addiction in America.* 1982. Cambridge, MA: Harvard University Press, 2001.

——. *Violent Land: Single Men and Social Disorder from the Frontier to the Inner City.* Cambridge, MA: Harvard University Press, 1996.

Courtwright, David T., and Timothy A. Hickman. "Modernity and Anti-modernity: Drug Policy and Political Culture in the United States and Europe in the Nineteenth and Twentieth Centuries." In *Drugs and Culture: Knowledge, Consumption, and Policy,* edited by Geoffrey Hunt, Maitena Milhet, and Henry Bergeron, 213–24. Burlington, VT: Ashgate, 2011.

Crosby, Alfred W. *America's Forgotten Pandemic: The Influenza of 1918.* New York: Cambridge University Press, 2003.

DeArment, Robert K. *Bat Masterson: The Man and the Legend.* Norman: University of Oklahoma Press, 1989.
Denton, Georgina. "'Neither Guns nor Bombs—Neither the State nor God—Will Stop Us from Fighting for Our Children': Motherhood and Protest in 1960s and 1970s America." *Sixties: A Journal of History, Politics, and Culture* 5 (December 2012): 205–28.
Duberman, Lucile. *Social Inequality: Class and Caste in America.* Philadelphia: J. B. Lippincott, 1976.
Duke, Steven B., and Albert C. Gross. *America's Longest War: Rethinking Our Tragic Crusade against Drugs.* New York: Tarcher Books, 1994.
Durkheim, Emile. *Suicide.* New York: Free Press, 1951.
Fairclough, Adam. *Race and Democracy: The Civil Rights Struggle in Louisiana, 1915–1972.* Athens: University of Georgia Press, 1995.
Gelman, Andrew, Lane Kenworthy, and You-Sung Su. "Income Inequality and Partisan Voting in the United States." *Social Science Quarterly* 91 (December 2010): 1203–19.
Ghatak, Saran. "The Opium Wars: The Biopolitics of Narcotic Control in the United States, 1914–1935." *Critical Criminology* 18 (March 2010): 41–56.
Glassner, Barry. *The Culture of Fear: Why Americans Are Afraid of the Wrong Things.* New York: Basic Books, 1999.
Goode, Erich, and Nachman Ben-Yehuda. "Moral Panics: Culture, Politics, and Social Construction." *Annual Review of Sociology* 20 (1994): 149–71.
———. *Moral Panics: The Social Construction of Deviance.* Oxford: Blackwell, 1994.
Greenfield, Robert. *Timothy Leary: A Biography.* New York: Houghton Mifflin, 2006.
Greer, Chris, ed. *Crime and Media: A Reader.* New York: Routledge, 2009.
Herdt, Gilbert. *Moral Panics, Sex Panics: Fear and the Fight over Sexual Rights.* New York: New York University Press, 2009.
Hickman, Timothy A. *The Secret Leprosy of Modern Days: Narcotic Addiction and Cultural Crisis in the United States, 1870–1920.* Amherst: University of Massachusetts Press, 2007.
Hier, Sean P. "Raves, Risks and the Ecstacy Panic: A Case Study in the Subversive Nature of Moral Regulation." *Canadian Journal of Sociology* 27 (Winter 2002): 33–57.
———. "Tightening the Focus: Moral Panic, Moral Regulation and Liberal Government." *British Journal of Sociology* 62, no. 3 (2011): 523–41.
"History of the Testor Corporation." Reference for Business: Company History Index. Accessed June 14, 2011. http://www.referenceforbusiness.com/history2/77/The-Testor-Corporation.html.
Hohenstein, Kurt. "Just What the Doctor Ordered: The Harrison Anti-narcotic Act, the Supreme Court, and the Federal Regulation of Medical Practice, 1915–1919." *Journal of Supreme Court History* 26, no. 3 (2001): 231–56.
Honey, Michael K. *Going down Jericho Road: The Memphis Strike, Martin Luther King's Last Campaign.* New York: W. W. Norton, 2007.
Horowitz, Daniel. *The Anxieties of Affluence: Critiques of American Consumer Culture, 1939–1979.* Amherst: University of Massachusetts Press, 2004.
Hunt, Arnold. "'Moral Panic' and Moral Language in the Media." *British Journal of Sociology* 48 (December 1997): 629–48.
Jenkins, Philip. *Intimate Enemies: Moral Panics in Contemporary Great Britain.* New York: Aldine De Gruyter, 1992.

———. *Moral Panic: Changing Concepts of the Child Molester in Modern America*. New Haven: Yale University Press, 1998.

———. *Synthetic Panics: The Symbolic Politics of Designer Drugs*. New York: New York University Press, 1999.

Jones, Marian Moser. "Poison Politics: A Contentious History of Consumer Protection against Dangerous Household Chemicals in the United States." *American Journal of Public Health* 103 (May 2013): 801–12.

Jones, Peter d'A. *The Consumer Society: A History of American Capitalism*. New York: Penguin, 1965.

Joseph, Peniel E. *Waiting 'til the Midnight Hour: A Narrative History of Black Power in America*. New York: Henry Holt, 2005.

Katsinas, Stephen G. "George C. Wallace and the Founding of Alabama's Public Two-Year Colleges." *Journal of Higher Education* 65 (July–August 1994): 447–72.

King County Bar Association. "Drugs and the Drug Laws: Historical and Cultural Contexts." *Report of the Legal Frameworks Group to the King County Bar Association Board of Trustees*. Seattle: King County Bar Association Drug Policy Project, 2005.

Kotarba, Joseph A., and Andrea Fontana, eds. *The Existential Self in Society*. Chicago: University of Chicago Press, 1984.

Kuzmarov, Jeremy. *The Myth of the Addicted Army: Vietnam and the Modern War on Drugs*. Amherst: University of Massachusetts Press, 2009.

Lasch, Christopher. *The Culture of Narcissism: American Life in an Age of Diminishing Expectations*. New York: W. W. Norton, 1978.

Lee, Martin A., and Bruce Shlain. *Acid Dreams: The CIA, LSD, and the Sixties Rebellion*. New York: Grove Press, 1986.

McEnany, Arthur E. *Membership in the Louisiana Senate, 1880–2004*. Baton Rouge: Louisiana State Senate, 2002.

Musto, David F. *The American Disease: Origins of Narcotic Control*. 1973. New York: Oxford University Press, 1987.

Noel, Thomas J. *Denver: Mining Camp to Metropolis*. Niwot: University Press of Colorado, 1990.

O'Connor, Denis J. *Glue Sniffing and Volatile Substance Abuse by Schoolchildren and Adolescents*. PhD diss. University of Newcastle-upon-Tyne, England, 1986.

Poynter, David R. *Membership in the Louisiana House of Representatives, 1812–2012*. Baton Rouge: Louisiana House of Representatives, 2010.

Rasmussen, Nicolas. *On Speed: The Many Lives of Amphetamine*. New York: New York University Press, 2008.

Reeves, Jimmie L., and Richard Campbell. *Cracked Coverage: Television News, the Anti-cocaine Crusade, and the Reagan Legacy*. Durham: Duke University Press, 1994.

Reinarman, Craig, and Harry G. Levine. *Crack in America: Demon Drugs and Social Justice*. Berkeley: University of California Press, 1997.

Reinarman, Craig, Dan Waldorf, Sheigla B. Murphy, and Harry G. Levine. "The Contingent Call of the Pipe: Bingeing and Addiction among Heavy Cocaine Smokers." In *Crack in America: Demon Drugs and Social Justice*, edited by Craig Reinarman and Harry G. Levine, 77–97. Berkeley: University of California Press, 1997.

Reiner, Robert. "Media Made Criminality: The Representation of Crime in the Mass Media." In *The Oxford Handbook of Criminology*, edited by Robert Reiner, Mike Maguire, and Rod Morgan, 302–40. New York: Oxford University Press, 2002.

Robins, Lee N. "Vietnam Veterans' Rapid Recovery from Heroin Addiction: A Fluke or Normal Expectation?" *Addiction* 88 (August 1993): 1041–54.
Rose, Max, and Frank R. Baumgartner. "Framing the Poor: Media Coverage and US Poverty Policy, 1960–2008." *Policy Studies Journal* 41 (February 2013): 22–53.
Rosides, Daniel W. *The American Class System: An Introduction to Social Stratification*. Boston: Houghton Mifflin, 1976.
Schissel, Bernard. *Blaming Children: Youth Crime, Model Panics, and the Politics of Hate*. Halifax, Nova Scotia: Fernwood Publishing, 1997.
Schneider, Eric C. *Smack: Heroin and the American City*. Philadelphia: University of Pennsylvania Press, 2008.
Scott, Jacqueline. "Is It a Different World to When You Were Growing Up? Generational Effects on Social Representations and Child-Rearing Values." *British Journal of Sociology* 51 (June 2000): 355–76.
Shaffer, Robert. "Public Employee Unionism: A Neglected Social Movement of the 1960s." *History Teacher* 44 (August 2011): 489–508.
"Silent Partners: The Role of the Church in Liberalizing Georgia's Abortion Laws." *Georgia Right to Life*. Accessed June 14, 2010. http://www.grtl.org/history.asp.
Simon, William, and John H. Gagnon. "The Anomie of Affluence: A Post-Mertonian Conception." *American Journal of Sociology* 82 (September 1976): 356–78.
Smith, Emma. "Failing Boys and Moral Panics: Perspectives on the Underachievement Debate." *British Journal of Educational Studies* 51 (September 2003): 282–95.
Spillane, Joseph F. *Cocaine: From Medical Marvel to Modern Menace in the United States, 1884–1920*. Baltimore: Johns Hopkins University Press, 2000.
———. "Federal Policy in the Post-Anslinger Era: A Guide to Sources, 1962–2001." In *Federal Drug Control: The Evolution of Policy and Practice*, edited by Jonathon Erlen and Joseph F. Spillane, 209–20. Binghamton, NY: Haworth Press, 2004.
———. "The Road to the Harrison Narcotics Act: Drugs and Their Control, 1875–1918." In *Federal Drug Control: The Evolution of Policy and Practice*, edited by Jonathon Erlen and Joseph F. Spillane, 1–24. Binghamton, NY: Haworth Press, 2004.
Springhall, John. *Youth, Popular Culture, and Moral Panics: Penny Gaffs to Gangsta Rap, 1830–1996*. New York: St. Martin's, 1999.
Ungar, Sheldon. "Moral Panic versus the Risk Society: The Implications of the Changing Sites of Social Anxiety." *British Journal of Sociology* 52 (June 2001): 271–91.
Victor, Jeffrey S. "Moral Panics and the Social Construction of Deviant Behavior: A Theory and Application to the Case of Ritual Child Abuse." *Sociological Perspectives* 41, no. 3 (1998): 541–65.
Weir, William. *In the Shadow of the Dope Fiend: America's War on Drugs*. North Haven, CT: Archon Books, 1995.
White, Cameron, John L. Oliffe, and Joan L. Bottorff. "From Promotion to Cessation: Masculinity, Race, and Style in the Consumption of Cigarettes, 1962–1972." *American Journal of Public Health* 103 (April 2013): 44–55.
Zinberg, Norman E. *Drug, Set, and Setting: The Basis for Controlled Intoxicant Use*. New Haven: Yale University Press, 1986.
Zinberg, Norman E., and John A. Robertson. *Drugs and the Public*. New York: Simon and Schuster, 1972.
Zumello, Christine. "The 'Everything Card' and Consumer Credit in the United States in the 1960s." *Business History Review* 85 (Autumn 2011): 551–75.

# INDEX

7-Eleven, 170

acetone, 28, 45, 46, 48, 65, 66, 110, 114, 118, 184, 185, 187, 190, 191, 193, 220n, 221n, 224n, 225n
Acker, Caroline Jean, 88, 107
Adams, Tom, 33–34
Adams, W. Thomas, 57–59, 60
Aichhorn, August, 205n
AIDS, 169, 177
*Airplane!*, 11
Alabama House of Representatives, 141–143
Alabama Senate, 142
alcohol, 18, 37, 48, 59, 60, 67, 68, 69, 70, 109, 113, 126, 135
Alex, Ted, 101
Alexander, Michelle, 11, 169, 176
Alexander, William B., 222n
Alexandria, Louisiana, 124
aliphatic hydrocarbons, 68, 114, 179, 181–182
Allegheny County coroner's office, 105
Allman, Leo M., 43
allylisothiocyanate, 119–120, 142, 156, 168
Altamonte Springs, Florida, 128, 221n
Ambroid Company, 74–75
*American*, 34–35
American Home Products, 167–168
American Medical Association, 68, 75, 88
American Psychoanalytic Association, 12, 24
American Social Health Association, 123
American Society of Humanistic Education, 79–80
amphetamine, 10, 47, 67, 97, 149, 150
amyl acetate, 38, 46, 185, 187, 191, 193, 220n, 221n, 224n
amyl alcohol, 191, 224n
amyl chloride, 191, 224n
Anaheim, California, 50, 52, 108
Anderson, John, 156

Anniston, Alabama, 141, 225n
Anti-Drug Abuse Act (1976), 169
Anti-Drug Abuse Act (1978), 169
Anzelmo, Salvador, 183, 219n
Applied Biological Sciences Laboratory, 74
Arkansas River, 26
Associated Community Teams of Harlem (ACT), 79
Associated Press, 141
Atkinson, Stuart, 81
Atlanta, Georgia, 134–136, 137, 138, 139, 140
*Atlanta Journal*, 134, 136, 137
Atlanta Police Department, 137, 138
Avro Lancaster, 34, 205n

Baltimore, Maryland, 42, 150, 151, 165
Bank, Bert, 141
barbiturates, 69, 80, 106, 149, 153
Barker, Gordon, 57–59, 60
Barthel, Christopher E., 162
Bartlett, Sylvan, 70–71
Baudelaire, Charles, 14
Becker, Howard, 6
Bedford-Stuyvesant, Brooklyn, 49
Beeson, James E., 183, 219n
Bell, Clyde F., Jr., 219n
Bell, John W., 221n
Belzoni, Mississippi, 132
Ben-Yehuda, Nachman, 5, 7
Benzedrine, 47, 75
benzene, 18, 19, 24, 29, 38, 46, 48, 63, 65, 99, 105, 110, 114, 117, 118, 179, 180–181, 185, 187, 191, 193, 203n, 220n, 221n, 224n, 225n
Berkeley, California, 87, 89, 148
Bernhard, William F., Dr., 183, 219n
Bexar County, Texas, 68
Bill's Rough Riders, 83
Biloxi, Mississippi, 132, 222n
Birmingham, Alabama, 141
*Birmingham News*, 141, 142
Black Diamonds, The, 3–4

246  INDEX

Black Panther Party, 152
Black Power, 126, 168
Bleuler, Eugen, 13
Blyer, Lee L., 52, 208n
Boesch, Edward L., 183, 219n
Bolton, Arthur K., 140
Bonaparte, Napoleon, 14
Bond, Julian, 224n
*Bond v. Floyd* (1956), 224n
Bordes, Charles, III, 183, 219n
Borsch, Henry, 123
Boston, Massachusetts, 173, 208n
Boyer, Paul, 52
Breed, Chum, 82
Bridgeport, Connecticut, 80, 81
Bridges, Lloyd, 11
Brill, Abraham Arden, 12–15, 24, 201n
Brill, Henry, 83, 155, 227n
*British Medical Journal*, 42, 105
Brooklyn, New York, 3, 6, 49, 54, 76, 148, 151
Brotman, Richard, 100–101, 103
Broxon, John R., 220n
Bruce, Lenny, 34, 177
Burnstein, Beatrice S., 77
butyl acetate, 38, 45, 46, 66, 114, 185, 187, 191, 193, 220n, 221n, 224n, 225n
butyl alcohol, 38, 46, 66, 114, 185, 188, 191, 193, 220n, 221n, 224n, 225n
butyl ether, 191, 224n

Caddo Parish, Louisiana, 124, 219n
Calcasieu Parish, Louisiana, 124, 219n
California Achievement Test, 58
Campbell, Tommy, 132
Canadian Commission of Inquiry into the Non-Medical Use of Drugs, 168
Capone, Al, 3
carbon disulfide, 18, 191, 224n
carbon tetrachloride, 28, 38, 46, 48, 119, 185, 191, 193, 220n, 221n, 224n, 225n
Carbona cleaning fluid, 76
Carlton, Frank, 222n
Carmody, John, 145
Carnegie Hall, 34
Carrell, Thomas, 72
Case Western Reserve University, 89

Casey, Joseph S., 183, 219n
Causey, Gordon E., 183, 219n
Centers for Disease Control, 56
Central Intelligence Agency, 28
Chafin, Clinton, 135, 137
Chaisson, Joel T., 183, 219n
Chapel, James, 116–117
Charleston, Mississippi, 133, 222n
Chicago, Illinois, 59–60, 89, 108, 109, 110, 158
Chicago Police Department, 59–60, 110
  Youth Division, 59–60
Child Protection Act, 85
Child Safety Protection Act, 229n
Chiles, Lawton, 128
chlorinated hydrocarbon, 50, 96, 179, 181, 203n
chloroform, 4, 18, 19, 28, 38, 46, 185, 188, 191, 193, 220n, 221n, 224n, 225n
Chomsky, Noam, 176
*Christian Science Monitor*, 159
Cincinnati, Ohio, 18, 108
Citizens United Organization Association, 151
Civil Defense Board, 80
Civil Rights Act of 1954, 122
Clacton, England, 5
Clark, Kenneth, 79
Clayton, Ralph R., 221n
cleaning fluid, 45, 48, 76, 155, 181
Clinger, Orris W., 19–21
Cobble Hill, Brooklyn, 3
cocaine, 13, 40, 43, 50, 68, 69, 75, 83, 84, 113, 127, 169, 170, 171, 206n
Cocoa Beach, Florida, 127, 220–221n
Cohen, Stanley, 5, 199n
Cole, Luther F., 183, 219n
Collegium Internationale Neuro-Psychopharmacologicum, 83, 84
Colorado Children's Code, 89
Colorado House of Representatives, 89
Colorado State Fair, 26
Columbia University, 70, 116, 117, 148, 166
  School of Public Health, 166
Colvin, Douglas Glenn, 171–172
Commission on Segregation, 89
Community Memorial Hospital, 203

Comprehensive Crime Control Act, 169
Consolidated Tenants League, 83
*Consumer Reports*, 40–41, 53
Consumer's Union, 40
Controlled Substances Act, 169
Cook County, Illinois, 89
Coodley, Alfred E., 31–32
Cooper, Joe Henry, 219n, 220n
Cooper, Robert, 47
Coram, Robert, 134, 135
Coreil, Joseph Emil, 183, 219n
Corliss, Leland M., 66, 212n
Corr, Philip, 52
Corr, William Ervin, 222n
cough syrup, 42, 106
Court of Special Appeals of Maryland, 165
Courtwright, David, 79, 107
Crais, Arthur A., 183, 219n
Crenshaw County, Alabama, 142
Crescent City Independent Voters League, 125
Cross, J. Emory, 220
Crowther, Deborah, 104, 107
Crowther, Kimberly, 104, 107
Culver, John, 141
Currigan, Thomas, 89
Curtis Park Recreation Center, 33
cyclohexanone, 38, 48, 184, 185, 188, 190, 193, 220n, 221n, 225n

Dade County Circuit Court, 130
Daley, Stephen K., 183, 219n
Darien, Connecticut, 80–81
Darien High School, 80
Davis, Richard, 98–99
Davis, Russell, 222n
Deeb, Richard J., 221n
DeKalb County, Georgia, 136, 137
DeLand, Florida, 128, 221n
Democratic National Convention, 148
Dempsey, Oberia D., 73–74
Denver, Colorado, 4, 8, 26, 27, 28, 29, 30, 33, 34, 35, 36, 37, 39, 41, 43, 53, 58, 61, 66, 73, 78, 79, 86, 87, 88, 89, 93, 101, 102, 103, 115, 116, 137, 143, 145, 159, 160, 166, 167, 195, 196, 198, 205n, 228n

Denver Glue-Sniffing Project, 101, 106, 107, 195–198
Denver International Airport, 87
Denver Juvenile Court, 33, 36, 78, 84, 87, 89, 101, 106, 159, 195
Denver Juvenile Hall, 61
Denver Police Department, 29, 30
  Vice Bureau, 73
*Denver Post*, 4, 28, 33, 38
Department of Justice, 143, 152
  Bureau of Narcotics and Dangerous Drugs, 143, 152
Department of the Treasury, 143, 152
DePaul University, 89
Detroit, Michigan, 104, 105, 108
diethylcarbonate, 191, 224n
diethylene oxide, 191, 224n
dipropyl ketone, 191, 224n
Dodds, Josiah, 61
Done, Alan K., 96–97, 107–110, 111, 112n,
*Dragnet*, 214n
Drug Abuse Control Act, 162
Drug Abuse Education Act (Florida), 130
*Drug Abuse: Escape to Nowhere*, 146–147
Drug Enforcement Agency, 169
Duke's Motel (Baltimore), 165
Du Pont Chemical, 75
Dupuy, Clarence O., 123, 124, 125
Durkheim, Emile, 10, 201n
Duro-Matic Products, 44
Dwyer, William J., 183, 219n

Early, Thomas A., Jr., 183, 219n
Easson, William M., 36, 37
East Baton Rouge Parish, Louisiana, 124, 219n
Eisenhower, Dwight D., 64, 89, 152
*The Electric Kool-Aid Acid Test*, 82
Ellis, Havelock, 12–13
Ellison, Willie S., 63, 65, 66
Ely, Dale F., 99
ether, 4, 18, 45, 50, 68, 154, 182, 191, 224n
ethyl acetate, 38, 46, 48, 114, 184, 185, 188, 190, 191, 193, 220n, 221n, 224n, 225n
ethyl alcohol, 38, 46, 48, 148, 185, 188, 191, 193, 220n, 221n, 224n, 225n
ethyl butyrate, 191, 224n

ethylene dichloride, 38, 46, 191, 203n, 224n
ethylene glycol monoethyl ether, 191, 224n
ethylene glycol monomethyl ether acetate, 191, 224n
Etowah, Alabama, 142

Facebook, 56
Faircloth, Earl, 130
Fairfax, Virginia, 146, 170, 172
Farrington, Palmer, 76
Featherstone, Harold G., 221n
Federal Bureau of Drug Abuse Control, 143, 148, 152
Federal Bureau of Investigation, 88, 100–101, 123
Federal Bureau of Narcotics, 6, 143, 152
Federal Hazardous Substances Act, 35, 36, 44, 85, 156, 161, 163, 229n
Ferenczi, Sándor, 15, 24
fingernail polish, 45
fingernail polish remover, 48, 76, 110, 115, 168
Fink, Olaf, 219n, 220n
Firestone, George, 127–128, 129, 221n
Fist, John, 82
Fite, Rankin, 142–143
Fliess, Wilhelm, 13, 24
*Florida Health Notes*, 130
Florida House of Representatives, 127–128, 129
Florida School for Boys, 154
Florida Senate, 127, 129
Florida Supreme Court, 130–131
Food and Drug Administration, 43–44, 143, 152, 208n
  Bureau of Drug Abuse Control, 143, 152
Fort, Joel, 40, 68, 82–83, 154, 205n
Fortier, Donald L., 183, 219n
Freedman, Alfred, 67
Freeman, Richard, 136
Freon, 87, 151, 167, 170, 180
Fresno, California, 74, 209n
Freud, Sigmund, 8, 13, 16, 24, 31, 117, 202n
Friedman, Paul, 24
Frost, Laurence L., 43
Fulton County, Georgia, 136
Fulton County Juvenile Court, 134

Gadsden, Alabama, 141, 225n
Gallen, Thomas M., 221
Garcia, Arthur, 73
Gardner, Jay, 138
Gartland, Philip, 29, 33
gasoline, 7, 9, 15–20, 21, 24, 32, 33, 36, 37, 39, 45, 46, 47, 48, 68, 70–71, 106, 110, 111, 118, 123, 154, 163, 175, 177, 179, 180, 181, 182, 205n, 207n
Gellman, Vera, 115
General Hospital of Fresno County (California), 209n
Georgia Department of Public Health, 136–137
Georgia House of Representatives, 139–140, 224n
Georgia Senate, 138–140, 190
Gerard, Donald, 21–22, 23n, 25n, 204n
Gilded Age, 40, 144
Gill, William A., Jr., 183, 219n
Gilliam, Philip, 36, 89
Gioscia, Victor, 97–98, 99
Glaser, Frederick, 69–70
Glaser, Helen, 37–39, 40, 41, 53, 61
Glaser, Lewis H., 47
Glassner, Barry, 144
Glendale, California, 74
Glue Angels, The, 151
Golden, Colorado, 33, 57
Gollott, Tommy, 222n
Gomma, Leo, 36
Goode, Erich, 5, 7
Gordon, Dorothy, 84
Gowanus Canal, 76
Graves, Charles Edward, 224n
Gray, Chester H., 51
Grayson, John William, 141
Great Migration, 39
Great Society, 78
Greece, 4
Gregson, Vernon J., 183, 219n
Gremillion, Allen C., 183, 219n
Grove Press, 177
Guidry, Richard P., 183, 219n

Haight-Ashbury, 73
hairspray, 155

Halbrook, David, 132
Hambridge, Gare, 15
Hanson, Robert, 95–96
Harlem, New York, 49, 74, 79, 80, 83, 84
Harlem Youth Opportunities Unlimited (HARYOU), 79
Harless, Michael, 167
Harrisburg, Pennsylvania, 88–89
Harrison County, Mississippi, 132, 222n
Harrison Narcotics Act, 88, 169, 207n
Hartmann, Heinz, 202n
HARYOU-ACT, 79, 83, 84
Hattiesburg, Mississippi, 132
Hawaii House of Representatives, 161
Hawaii Senate, 161
Hayden, Tom, 10
Hazardous Products Act (Canada), 168
Hazardous Products (Hazardous Substances) Regulations (Canada), 168
Health Research Council (New York), 152
Helrod, Robert H., 221n
Hempstead, New York, 76
Henderson, Warren S., 221n
Henley, Marvin B., 222n
Henry IV, 14
heroin, 7, 22, 24, 32, 39, 53, 66, 67, 68, 69, 74, 75, 79, 81, 83, 84, 97, 105, 106, 113, 121, 123, 126, 152, 153, 157, 162, 170, 171, 211n
Herrin, Ralph, 222n
Hersey, John, 82
Hessler, Ernest J., Jr., 183, 219n
hexane, 38, 46, 48, 66, 114, 179, 180, 184, 185, 188, 190, 193, 220n, 221n, 225n
High Royds Hospital (Leeds), 119
Higher Education Act, 161
Hillensbeck, Harry J., 183, 219n
Hinds County, Mississippi, 132, 222n
Hobby Industry Association of America, 8, 9, 36, 44, 45, 47, 50, 51–52, 54, 65, 74, 75, 88, 98–99, 102, 107, 119, 162, 163, 164, 167
Glue Sniffing Committee, 47
Hocking, Abe, 80
Hoffman, Abbie, 148
Hoffman, Thomas Marx, 183, 219n
Hollahan, George L., Jr., 220n
Hollins, Harry M., 219n, 220n
Holt, Elmo, 134–136

Houston, Texas, 108, 136
Hull, Ellis F., 124–125

Illinois Department of Public Health, 110
Imperial Beach, California, 72
Industrial Revolution, 40
Influenza epidemic of 1908, 173–174
Interfaith Neighbors, 151
isobutyl alcohol, 191, 224n
isopropanol, 48, 184, 185, 188, 190, 193, 220n, 221n, 225n
isopropyl acetate, 66, 185, 188, 191, 193, 220n, 221n, 225n
isopropyl alcohol, 28, 38, 46, 66, 114, 185, 188, 191, 193, 220n, 221n, 224n, 225n
isopropyl ether, 191, 224n
Iverson, William, 83

Jackson, Alabama 151, 225n
Jackson, Mississippi, 132, 222n
Jacksonville, Florida, 128, 220n
Jaffe, Jerome H., 211
James, George, 49
Jarman, John, 164
Jefferson Parish, Louisiana, 124, 219n
Jenkins, Philip, 144
Jewish Family and Children's Service, 97
Jim Crow, 10, 176
Johns Hopkins Hospital, 42
Johnson, Elizabeth J., 127–128, 129, 220–221n
Johnson, Lyndon B., 64, 78
Johnson, Nelson A., 19, 21
Johnson, Samuel, 28
Johnson, Virginia, 177
Johnston, J. Bennett, Jr., 183, 219n
Jones, Harold, 33
Jones, William Valentine, 222n
Jonson, Ben, 14
*Journal of the American Medical Association*, 30, 37, 41, 53
*Journal of Pediatrics*, 66
*Journal of School Health*, 62
*Juvenile Court Judges Journal*, 106

Karlson, Axel, 44
Keith, Hobart, 153

Keniston, Kenneth, 10
Kennedy, John F., 64
Kent State University, 148
Keogh, Joe, 183, 219n
kerosene, 15, 16, 46, 182
Kesey, Ken, 72, 149
King, Martin Luther, 152
Kinsey, Alfred, 177
Klein, Ben, 35, 36
Koch-Grüngerg, Theodor, 201n
Kornetsky, Conan, 21–22, 23, 25, 204n
Kouri, Joseph, 54
Krafft-Ebing, Richard von, 13, 24
Kupperstein, Lenore R., 115–116
Kutner, Murray, 146

lacquer thinner, 48, 96, 148, 181, 182
Lacy, J.L., 183, 219n
La Honda, California, 72
Lafayette, Louisiana, 124
Lake Charles, Louisiana, 124
Lake Ponchartrain, 121, 122
Lakeland, Florida, 128
Lambeth Hospital, 49
*The Lancet*, 105
Land, Herman, 155
Landau, Jack, 76, 83
Landers, Ann, 153, 166
Lane, David C., 221n
Lardner, George, Jr., 76–77, 84
Lasch, Christopher, 200–201
Lauricella, Francis E. "Hank," 219n
League for Spiritual Discovery, 85
League of Nations, 26
Leary, Timothy, 81, 85, 88, 101, 149
LeBleu, Conway, 219n, 220n
LeBreton, Edward F., Jr., 183, 219n
Lederberg, Joshua, 156, 164
*Lehrbuch de Psychiatrie*, 13
Lennon, Robert L., 132
Lepage's Liquid Solder, 65
Levine, Harry G., 176
Lexington, Kentucky, 21, 69
lighter fluid, 4, 48, 68
Lincoln, James, 104
Lindesmith, Alfred R., 65, 67

Lohman, Joseph, 89–93, 95, 96, 97, 103, 107, 116
Long Island, New York, 42, 81, 97, 146
Lookout Mountain School for Boys, 33, 57
Los Angeles, California, 30, 31, 44, 73, 109, 110, 141, 209n
Los Angeles County General Hospital, 62
Los Angeles County Probation Office Juvenile Hall, 59
Louisiana House of Representatives, 123–125, 127, 183, 219n
Louisiana Senate, 123–125, 127, 140, 185, 220n
*Louisiana v. Dimopoullas* (1962), 127, 130
Love, John, 89
Low, Robert A., 49
Luger, Milton, 99
LSD (Lysergic acid diethylamide), 9, 28, 82, 85, 101, 111, 114, 121, 123, 146, 148, 149, 152, 153, 157, 162, 170, 208n, 214n

Machle, William, 18
MacMillan, William L., 74
Marcel, Cleveland J., Sr., 183, 219n
Marcus, Walter F., Jr., 124
Marianna, Florida, 154
Marihuana Tax Act, 6, 169
marijuana, 6, 7, 10, 11, 16, 53, 68, 73, 74, 76, 80, 81, 82, 85, 100, 101, 105, 106, 123, 146, 149, 150, 152, 153, 157, 162, 164, 166, 167, 169, 170, 171, 176
Marks, George, 150
Marks, Sutton, 222n
Maryland State Health Department, 150
Massachusetts Department of Public Health, 41
Massengale, Oliver, 37–39, 40, 41, 53, 61
Masters, William, 177
Masterson, Bat, 26
Mathews, John E., Jr., 220n
Maugham, Somerset, 14
Maui High School, 161
McClain, Joseph A., Jr., 221n
McCready, George, 34, 206n
McDade, Helen, 222n

McKinley, William E. ("Bill"; Mississippi), 132, 222n
McMyne, William, 80
Meloff, William, 93–96
Meriden School for Boys, 81
Merry, Julius, 106, 216–217n
Merry Pranksters, 72
mescaline, 82, 111, 114, 149
Methadrine, 150
methyl acetate, 224n
methyl alcohol, 48, 191
methyl amyl acetate, 191
methyl amyl alcohol, 191, 224n
methyl benzene, 18, 105, 180
methyl ethyl ketone, 38, 46, 48, 114, 184, 185, 188, 190, 193, 220n, 221n, 225n
methyl isobutyl ketone, 38, 46, 48, 184, 185, 188, 190, 191, 193, 220n, 221n, 224n, 225n
methylcellosolve acetate, 38, 48, 184, 185, 188, 190, 191, 193, 220n, 221n, 224n, 225n
methylisobutyl ketone, 38, 46, 48, 184, 185, 188, 190, 191, 193, 220n, 221n, 224n, 225n
Metropolitan Crime Commission (New Orleans), 124
Miami, Florida, 47, 127, 128, 220n
Millbrook, 88, 101, 103
Milledgeville, Georgia, 137
Miller, Charles D., 44, 45, 75, 156
Miller, Roderick L., 183, 219n
Mink, John Francis, 161
Mink, Patsy (Takemoto), 9, 10, 161–165, 166
Mississippi Department of Public Welfare, 131
Mississippi House of Representatives, 131–133, 189, 222n
Mississippi Senate, 131–133, 222n
Mitchell, Charles, 222n
Mobile, Alabama, 141
Mollere, Jules G., 219–220n
Moomaw, Joe, 29, 33
Monroe, Louisiana, 123
Montana State Vocational School for Girls, 76
Monterey Park, California, 152–153
Morgan, S.M., Jr., 183, 219n
Morrison, Colorado, 57

Mosby, Joseph McRae, 222n
Moss, Joseph, 222n
mouthwash, 155
Mountview School for Girls, 57
Muirhead, Jean, 132, 222n
Munroe, John Thomas ("Tommy"), 132, 133, 222n
Murphy, Michael J., 75

naphthene (naphtha), 17, 46, 48, 68, 179, 181
Nassau County Board of Elections, 76
Nassau County District Court, 77
Nassau County Health Department, 42–43
Nassau County Narcotics Task Force, 76
National Association of Secondary School Principals, 153–154
National Cotton Council, 222n
National Education Association, 146–147
National Institute of Mental Health, 155
National Institute on Drug Abuse (NIDA), 170–171, 179, 230n
National Planning Commission for Washington, DC, 89
New Orleans, Louisiana, 121, 122–125, 219n, 220n
New Orleans Health Department, 124
New Orleans Voters Association, 125
New Orleans Voters League, 125
New York, New York, 3, 5, 22, 24, 34, 39, 42, 43, 47, 50, 51, 53, 54, 75, 76, 77, 80, 97, 98, 99, 101, 102, 108, 109, 110, 115, 145, 152, 208n
New York Academy of Medicine, 12
New York City Board of Health, 54
New York City Health Code, 98
New York City Health Department, 49, 98
New York City Poison Control, 78
New York Medical College, 67, 100
  Department of Psychiatry, 67, 100
New York Police Department, 49, 75, 79
  Youth Division, 49, 98
  Youth Investigation Unit, 98
New York State Division of Youth, 99
*New York Times*, 6, 43, 77, 82n, 151
New York University, 22
Newark, New Jersey, 47, 83, 84, 108

*Newsweek*, 47–49
Nietzsche, Friedrich, 14
nitrous oxide, 4, 7
Nixon, Richard, 10, 11, 64, 73, 143, 169, 176, 211n
Nobel Prize in Medicine, 164–165
Norman, Connolly, 13, 24
Northport, New York, 81
"Now I Wanna Sniff Some Glue," 172
Nunez, Samuel B., Jr., 183, 219n
nutmeg, 106, 111, 205n

O'Brien, Eugene G., 183, 219n
O'Brien, Frederick, 201n
O'Connor, Alicia R., 76, 77
O'Hearn, Taylor W., 183, 219n
O'Keefe, Michael, 219–220n
Oakland, California, 152
Occupational Health Service (Georgia), 137
olefins, 17, 181
Orleans Parish, Louisiana, 124, 125, 219n, 220n
Orleans Parish Progressive Voters League, 125
Orozco, Alfred, 152–153
Ottawa Civic Hospital, 118
Owens, Emmett, 222n
Owens, W.E., Jr., 225n
Owensboro, Kentucky, 170, 171, 172
Oxnard, California, 72
Oyster Bay, New York, 76

Pacific Ocean, 161
Paia Maui, Hawaii, 161
paint thinner, 4, 68, 69, 112n, 168, 181, 182
Palmer, Richard, 74
Pam cooking spray, 9, 167–168
paraffins, 17, 181
Parker, Larry, 219n, 220n
Parsippany, New Jersey, 74
Pasteur, Louis, 34, 178
Patridge, Corbet Lee, 222
PCP (phencyclidine), 230n
Pennsylvania State University, 161
Pensacola Naval Air Training Command, 52
*People v. Orozco* (1958), 152–153
Peterson, Robert, 155

Philadelphia, Pennsylvania, 76, 101–102
Philadelphia City Council, 76
Phillips, Donelan J., 83
*The Picture of Dorian Gray*, 14
Pinellas Park, Florida, 128, 221n
Plante, Kenneth, 221n
Playboys, The, 3–4
Poe, Edgar Allen, 14
Poison Prevention Packaging Act, 166
Port Huron Statement, 10
Portsmouth, Virginia, 151
Poston, Ralph R., 220n
Powars, Darleen, 62–63, 65
Powell, Adam Clayton, 79
Powell, John William, 222n
President's Advisory Commission on Narcotics, 53
Press, Edward, 66–67, 109n, 110, 112n
Priestly, Joseph, 4, 7
Prince Georges County, Virginia, 52
propylene dichloride, 191, 224n
propylene oxide, 191, 224n
Proust, Marcel, 24
psilocybin, 111, 149
Public Health Service, 44, 47
Pueblo, Colorado, 4, 26–28, 29, 33, 41, 58
Purtell, Thelma C., 84

Queens, New York, 171, 172

Rabin, Harry, 78
Radó, Sándor, 16–17
Ramone, Dee Dee, 171
Ramone, Joey, 172
The Ramones, 171–172
Rapides Parish, Louisiana, 124, 219n
Rasmussen, Nicolas, 67
Rawlin, John William, 97
Reagan, Ronald, 11, 73, 143, 169
Rebel, Inc., 47
Red Hook, Brooklyn, 3
Reddin, Thomas, 141
Redl, Fritz, 30
Reeves, R.B., 222n
Reinarman, Craig, 176, 211n
Research Center for Human Relations, 22
Revell Cement, 65

Rich, Dorothy, 147
Richardson, W.J., 183, 219n
Richelieu, Cardinal Armand Jean du Plessis, 15
Richmond, Virginia, 83
Rikers Island Prison, 76, 137
Robertson, Kenneth Barkley, 222n
Robins, Lee, 211n
Rockford, Illinois, 44, 156
Rome, Georgia, 139, 224n
Roszak, Theodore, 148
Rothstein, Ted L., 146
Rubin, H. Ted, 88–89, 102–103, 158–160
Rust, Robert W., 221n
Rutgers Symposium on Drug Abuse, 227
Ryan, Ruth, 145–146

Salt Lake City, Utah, 5, 45, 108, 110, 159
Salt Lake County General Hospital, 45
San Antonio, Texas, 15, 108
San Diego, California, 72
San Dimas, California, 72
San Fernando, California, 72
San Francisco, California, 35, 73, 208n
Santa Clara County Juvenile Probation Department, 63
Santostefano, Sebastiano, 61
Sapir, Eddie, 123, 124, 183, 219n
Savannah, Georgia, 138
Scanlon, John, 89
"The Scent of Danger," 45–47, 88
Schmideberg, Melitta, 31
Schneck, J.M., 19
Schneider, Eric C., 39
Scottsdale, Arizona, 30
Scudder, Gilbert H., 81
Secretary of Health, Education, and Welfare, 161, 163
Sessions, H.K., 137
Shakespeare, William, 30
Shanholtz, Mack I., 118
Shelley, Percy Bysche, 14
Shevin, Robert L., 220n
Shreveport, Louisiana, 124
Simon, Warren J., 183, 219n
Simpson, James C., 222n
Single Convention on Narcotic Drugs, 207n

Smith, James P., 183, 219n
Smith, Kline and French Laboratories, 146–147
Smith, Theodore, 222n
Smither, Charles, 183, 219n
Snickers, 9
Sobolevitch, Robert, 101–102
Social Security, 64
Social Welfare Planning Council, 122–124
Sokol, Jacob, 59, 63, 65, 209n
Soto, Domingo, 148
South Hill, Virginia, 203n
Sowell, Ralph, 222n
Special Action Office of Drug Abuse Prevention, 211n
Spicola, Guy W., 221n
Spillane, Joseph, 50
Springhall, John, 7
St. Thomas's Hospital (London), 105–106
Stamford, Connecticut, 80
Stanford University, 28
Staples High School, 150
Starnes, Richard, 139, 224n
Stefansson, Vilhjalmur, 201n
Sterling, James W., 59–60
Steve Allen Show, 34
Stone, Ben Harry, 222n
Stone, Richard B., 220n
Street Club Project, New York City, 47
Strider, H.C., 133, 222n
Strother, T.J., 183, 219n
Students for a Democratic Society, 10, 148
Sullivan, John P., 183, 219n
Sunbelt, 39, 128, 143
Susman, Ralph M., 115–116

Talbot, Richard E., 183, 219n
Tallahatchie County, Mississippi, 133
Tampa, Florida, 128, 220n, 221n
Tapia, Fernando, 70–71
Tausk, Viktor, 202n
Taylor, Daniel, 116–117
Tessier, George D., 125
Testor Chemical Company, 44
Testor Corporation, 9, 44, 45, 47, 50, 66, 75, 78, 119–120, 142, 156, 162, 164, 167, 168, 174, 175

Testor, Nils F., 44
Testor's Polystyrene Plastic Cement, 65
tetraethyl lead, 18
Thomas, Jerry, 220n
Thomas, Lester G., 106–107, 195, 217n
Thompson, Kenneth, 174
Tillman, Curtis, 136
*Time* magazine, 6, 45, 47, 49, 104
Tobriner, Walter, 52
Todd, John, 119
Tolstoy, Leo, 14
toluene, 5, 18–19, 24, 28, 38, 42, 44, 45, 46, 47, 48, 61, 65, 66, 73, 75, 87, 96–97, 99, 105, 106, 110, 111–113, 117–118, 119, 148, 153, 170, 179, 180, 184, 185, 188, 190, 191, 193, 220n, 221n, 224n, 225n
*Too Far to Walk*, 82
Trachtenburg, David, 155
Triche, Risley C., 183, 219n
trichlorochthane, 184, 190, 220n
trichloroethylene, 46, 185, 188, 191, 194, 220n, 221n, 224n, 225n
tricresyl phosphate, 38, 66, 188, 194, 221n, 225n
triorthocresyl phosphate, 180
True, James B., 222n
Truman, Harry S., 89
Tucson, Arizona, 4, 27–28, 29, 143
Tureaud, Alexander, 125
Turner, Alton, 142
turpentine, 15, 16, 17, 68, 207
Tuscaloosa, Alabama, 141

U.S. Department of Health, Education, and Welfare, 87–88, 161, 163
　Office of Juvenile Delinquency and Youth Development, 87–88, 116
　Center for Chronic Disease Control, 146
U.S. House of Representatives, 156, 161, 165
　Interstate and Foreign Commerce Committee, 162
U.S. National Guard, 148
U.S. Navy, 52, 53
U.S. Public Health Service Hospital, 21
Unified School District of Long Beach, California, 99
United Voters League, 124
University College Cork, 42
University Hospital, Saskatchewan, 36
University of Alberta, 93
University of California at Berkeley, 72, 148
　School of Criminology, 89
University of Chicago, 89, 161
University of Colorado Medical Center, 37, 61
　Children's Psychiatric Clinic, 89
University of Denver, 89
University of Hawaii, 161
University of Illinois Hospital, 158
University of Minnesota, 21, 32
University of Nebraska, 161
University of North Carolina at Chapel Hill, 114, 137
University of Puerto Rico, 116
University of Utah, 96
Upper Fifth Avenue Baptist Church, 74

Valdez, David, 152
Vance, Michael, 54
varnish, 4, 181
Vesich, Anthony J., Jr., 183, 219n
Vietnam, 10, 67, 68, 137, 148, 149, 152, 168, 176, 211n
Voting Rights Act of 1955, 121, 122

Wagner, Robert F., 53
Walker, Lillian W., 183, 219n
*Wall Street Journal*, 148
Wallace, George, 142
Wallace, Theodore, 81
War on Drugs, 11, 143, 169, 176, 177
Ward, Clara (Princesse de Caraman-Chimay), 14
Ward, James Joseph, 165
Warren, Pennsylvania, 19
Warren State Hospital, 19, 20
Washington, DC, 43, 51, 53, 54, 76, 79, 84, 89, 108, 146, 152, 170
Washington Board of Commissioners, 52–53
Washington Juvenile Court, 43
Washington Police Department, 51–52
　Juvenile Bureau, 51
*Washington Post*, 43, 44, 76, 145, 147, 156

Watson, Thomas Arnie, 222n
West Side High School, 83
Westport, Connecticut, 80, 150, 151
Weymouth, Massachusetts, 74
Wharton, New Jersey, 80
*What You Can Do about Drugs and Your Child*, 155
Whelton, Michael J., 42
Wilde, Oscar, 14
Wilson, Ethel, 67
Wilson, Harold S., 221n
Wilson, Rex, 61
Wilson, Woodrow, 26
Windermere, Florida, 128, 221n
Winek, Charles, 117–118
Winick, Charles, 123
Winnipeg, Manitoba, 114–115
Winston, Ellen, 84
Winters, John E., 51, 76
*The Winter's Tale*, 30
Wolfe, Tom, 82

Women's Educational Equity Act, 161
Woodfield, Clyde, 222n
Woolworth's, 44
World Health Organization, 65
World War II, 64, 126, 152
World's Fair (1954), 72
Wrenn, Robert, 51
Wright, Charles, 142–143

xylene, 18, 28, 38, 46, 114, 179, 181, 185, 220n

Yazoo City, Mississippi, 132
The Young Angels, 151
Young, C.W. Bill, 221n
Youth Aid Division of the Metropolitan Police, 52
Youth Development Center of South Philadelphia, 101

Zola, Emile, 14, 24

www.ingramcontent.com/pod-product-compliance
Lightning Source LLC
Chambersburg PA
CBHW020114010526
44115CB00008B/820